Synthesis Lectures on Artificial Intelligence and Machine Learning

Series Editors

Ron Brachman, Jacobs Technion-Cornell Institute, New York, USA

Francesca Rossi, Thomas J. Watson Research Center, IBM Research AI, Yorktown Heights, NY, USA

Peter Stone, University of Texas at Austin, Austin, TX, USA

This series publishes short books on research and development in artificial intelligence and machine learning for an audience of researchers, developers, and advanced students.

Aaron M. Roth · Dinesh Manocha ·
Ram D. Sriram · Elham Tabassi

Explainable and Interpretable Reinforcement Learning for Robotics

Aaron M. Roth
Department of Computer Science
University of Maryland
College Park, USA

Dinesh Manocha
Department of Computer Science
and Electrical and Computer Engineering
University of Maryland
College Park, MD, USA

Ram D. Sriram
National Institute of Standards and Technology
Gaithersburg, MD, USA

Elham Tabassi
National Institute of Standards and Technology
Gaithersburg, MD, USA

ISSN 1939-4608   ISSN 1939-4616 (electronic)
Synthesis Lectures on Artificial Intelligence and Machine Learning
ISBN 978-3-031-47517-7   ISBN 978-3-031-47518-4 (eBook)
https://doi.org/10.1007/978-3-031-47518-4

This is a U.S. government work and not under copyright protection in the U.S.; foreign copyright protection may apply 2024

All rights are solely and exclusively licensed by the Publisher, whether the whole or part of the material is concerned, specifically the rights of translation, reprinting, reuse of illustrations, recitation, broadcasting, reproduction on microfilms or in any other physical way, and transmission or information storage and retrieval, electronic adaptation, computer software, or by similar or dissimilar methodology now known or hereafter developed.
The use of general descriptive names, registered names, trademarks, service marks, etc. in this publication does not imply, even in the absence of a specific statement, that such names are exempt from the relevant protective laws and regulations and therefore free for general use.
The publisher, the authors, and the editors are safe to assume that the advice and information in this book are believed to be true and accurate at the date of publication. Neither the publisher nor the authors or the editors give a warranty, expressed or implied, with respect to the material contained herein or for any errors or omissions that may have been made. The publisher remains neutral with regard to jurisdictional claims in published maps and institutional affiliations.

This Springer imprint is published by the registered company Springer Nature Switzerland AG
The registered company address is: Gewerbestrasse 11, 6330 Cham, Switzerland

Paper in this product is recyclable.

Dedicated to all those who toil in pursuit of knowledge.

Acknowledgements

There are multiple people who helped make this book possible the authors would like to thank.

This work was funded by the National Institute of Standards and Technology (NIST), and multiple individuals from NIST reviewed this manuscript during its development.

We would further like to thank the reviewers from Morgan Claypool/Springer Nature who gave feedback that helped improve our work.

We would also like to thank the researchers who gave permission for images they created for other publications to appear in this book and others who helped obtain such permissions. They include Jieliang Luo, Nir Baram, Petra Poklukar, Martina Lippi, Maximilian Sieb, Miguel Jaques, Michael Burke, Sandy Huang, Utsav Patel, Shuby Deshpande, Bejanmin Eysenbach, Mathieu Seurin, Eadom Dessalene, and Dylan P. Losey.

Certain commercial software systems may be identified in this paper. Such identification does not imply recommendation or endorsement by the National Institute of Standards and Technology (NIST) or by the organizations of the authors; nor does it imply that the products identified are necessarily the best available for the purpose. Further, any opinions, findings, conclusions, or recommendations expressed in this material are those of the authors and do not necessarily reflect the views of NIST or any other supporting U.S. government or corporate organizations.

Contents

1	**Introduction**		1
	1.1 Motivation		1
	1.2 Background		6
		1.2.1 Reinforcement Learning	6
		1.2.2 Reinforcement Learning Versus Supervised, Unsupervised, and Imitation Learning	7
		1.2.3 Explainable Artificial Intelligence (XAI) and Explainable Reinforcement Learning (XRL)	8
		1.2.4 Explainable Robotics (X-Robotics)	8
	1.3 Selection Criteria		9
2	**Classification System**		11
	2.1 Prior Surveys on XRL or X-Robotics		11
	2.2 Existing Classification Terminology for XRL or X-Robotics		13
	2.3 The Attributes of our Classification System		16
		2.3.1 Hard Attributes	17
		2.3.2 Soft Attributes (General)	20
		2.3.3 Soft Attributes (Robot-Specific)	21
3	**Explainable Methods Organized by Category**		23
	3.1 Decision Tree		23
		3.1.1 Single Decision Tree	30
		3.1.2 Single Altered Decision Tree	32
		3.1.3 Multiple or Combined Decision Trees	32
	3.2 Saliency Maps		33
		3.2.1 Post-Hoc Saliency Maps via Backpropagation	33
		3.2.2 Intrinsic Saliency Maps	34
		3.2.3 Post-Hoc Saliency Maps via Input Perturbation	35
	3.3 Counterfactuals/Counterexamples		36
		3.3.1 Counterfactual by Input Perturbation or Extra Information	36
		3.3.2 Counterfactual by Model Checking	37

	3.4		State Transformation	39
		3.4.1	Dimension Reduction	39
		3.4.2	Meaningful Representation Learning	40
	3.5		Observation Based Methods	44
		3.5.1	Observation Analysis: Frequency or Statistical Techniques for Policy Understanding	45
		3.5.2	Observation Analysis: Human Communicative Trajectories for Goal Understanding	46
		3.5.3	Observation Analysis: A/B Testing	47
		3.5.4	Training Data Observation Analysis	47
		3.5.5	Interrogative Observation Analysis	48
	3.6		Custom Domain Language	49
	3.7		Constrained Learning	49
	3.8		Constrained Execution	50
	3.9		Hierarchical	52
		3.9.1	Hierarchical Skills or Goals	52
		3.9.2	Primitive Generation	54
	3.10		Machine-to-Human Templates	55
		3.10.1	Model-to-Text or Policy-to-Text Templates	55
		3.10.2	Query-Based NLP Templates	56
	3.11		Model Reconciliation	56
		3.11.1	Certain Model Reconciliation	56
		3.11.2	Uncertain Model Reconciliation	57
	3.12		Causal Methods	58
	3.13		Reward Decomposition	59
		3.13.1	Standard Reward Decomposition Methods	59
		3.13.2	Model Uncertainty Reward Decomposition	60
	3.14		Visualizations	61
	3.15		Instruction Following	62
	3.16		Symbolic Methods	64
		3.16.1	Symbolic Transformation	64
		3.16.2	Symbolic Reward	65
		3.16.3	Symbolic Learning	65
	3.17		Legibility or Readability	66
4	**Key Considerations and Resources**			**69**
	4.1		General Discussion	69
	4.2		Limitations of Some Methods in this Survey	71
	4.3		Human-Robot Interaction Considerations	72
	4.4		Legibility and Readability	74
	4.5		AI Safety	74
	4.6		Environments	75

5	**Opportunities, Challenges, and Future Directions**	77
	5.1 Opportunities, Challenges, and Future Directions	77
	5.2 Conclusion	88

References 89

Index 109

About the Authors

Aaron M. Roth is currently Head of Autonomy Technology for Black Sea. Previously, Aaron worked as a Research Scientist in the Distributed Autonomous Systems Group in the Navy Center for Applied Research in Artificial Intelligence at the Naval Research Laboratory and as a Researcher and Ph.D. student at the University of Maryland in the GAMMA lab. In both capacities, he led projects conducting research into autonomous robots and artificial intelligence, with specific focus on reinforcement learning, explainable/interpretable artificial intelligence, AI Safety, robot navigation, and human-robot interaction. He has also worked at several technology startups, in industries spanning healthcare, finance, and mobile consumer applications. Aaron's published research has been reported on by news publications worldwide, and he has presented his research speaking at international conferences in the academic research community and given talks explaining science, robotics, and artificial intelligence to the general public. He is the creator of and contributor to multiple open-source software projects. He holds a B.S.E. in Electrical Engineering from University of Pennsylvania and an M.S. in Robotics from Carnegie Mellon University. Aaron was Conference Chair for the Third Annual Artificial Intelligence Safety Unconference in 2021, an international conference about AI Safety. In both paid and volunteer capacities, Aaron has provided technical mentorship to undergraduate students, graduate students, and industry career professionals across the United States, Europe, and Australia. He is also a published science fiction author. Aaron has consulted on artificial intelligence topics for creative professionals including science fiction authors and game designers. Learn more at www.aaronmroth.com.

Dinesh Manocha is a Distinguished University Professor at the University of Maryland. He is also the Paul Chrisman Iribe Professor of Computer Science and Electrical and Computer Engineering. He is also the Phi Delta Theta/Matthew Mason Distinguished Professor Emeritus of Computer Science at Chapel Hill University of North Carolina. Mancha's research focuses on AI, robotics, computer graphics, augmented/virtual reality, and scientific computing, and has published more than 750 papers. He has supervised 48 Ph.D. dissertations, and his group has won 21 best paper awards at leading conferences. His group has developed many widely used software systems (with 2M+ downloads) and licensed them to more than 60 commercial vendors. He is an inventor of 17 patents, several of which have been licensed to industry. A Fellow of AAAI, AAAS, ACM, IEEE, NAI, and Sloan Foundation, Manocha is a ACM SIGGRAPH Academy Class member and Bézier Award recipient from Solid Modeling Association. He received the Distinguished Alumni Award from IIT Delhi and the Distinguished Career in Computer Science Award from Washington Academy of Sciences. He was also the co-founder of Impulsonic, a developer of physics-based audio simulation technologies, which Valve Inc acquired in November 2016. He is also a co-founder of Inception Robotics, Inc. Learn more at: http://www.cs.umd.edu/~dm

Ram D. Sriram is currently the chief of the Software and Systems Division, Information Technology Laboratory, at the National Institute of Standards and Technology. Before joining the Software and Systems Division, Sriram was the leader of the Design and Process group in the Manufacturing Systems Integration Division, Manufacturing Engineering Laboratory, where he conducted research on standards for interoperability of computer-aided design systems. Prior to joining NIST, he was on the engineering faculty (1986–1994) at the Massachusetts Institute of Technology (MIT) and was instrumental in setting up the Intelligent Engineering Systems Laboratory. Sriram has co-authored or authored more than 275 publications, including several books. Sriram was a founding co-editor of the *International Journal for AI in Engineering*. Sriram received several awards including: an NSFs Presidential Young Investigator Award (1989); ASME Design Automation Award (2011); ASME CIE Distinguished Service Award (2014); the Washington Academy of Sciences Distinguished Career in Engineering Sciences Award (2015); ASME CIE divisions Lifetime Achievement Award (2016); CMU CEE Lt. Col. Christopher Raible Distinguished Public Service Award (2018); IIT Madras Distinguished Alumni Award (2021). Sriram is a Fellow of AAIA, AIBME, ASME, AAAS, IEEE, INCOSE, SMA, and Washington Academy of Sciences, a Distinguished Member (life) of ACM and Senior Member (life) AAAI. Sriram has a B.Tech. from IIT, Madras, India, and an M.S. and a Ph.D. from Carnegie Mellon University, Pittsburgh, USA.

Elham Tabassi is the Chief of Staff in the Information Technology Laboratory (ITL) at the National Institute of Standards and Technology (NIST). She leads the NIST Trustworthy and Responsible AI program, which aims to cultivate trust in the design, development, and use of AI technologies by improving measurement science, standards, and related tools to enhance economic security and improve quality of life. She has been working on various machine learning and computer vision research projects with applications in biometrics evaluation and standards since she joined NIST in 1999. She is a member of the National AI Resource Research Task Force, a senior member of IEEE, and a fellow of Washington Academy of Sciences.

Introduction

Will robots inherit the earth? Yes, but they will be our children.

Marvin Minsky

1.1 Motivation

In the modern developed world, robots are present in daily life in myriad ways. Yet, the modern field of robotics is only decades old.

The first robots were designed and used in the 1950s and 1960s [247]. Researchers developed robots with tactile sensors, cameras, and movement capabilities. Stanford's SHAKEY robot could combine logical reasoning and physical action to plan navigation through rooms, flip light switches, and push objects in response to commands or based on its own plan [157]. By the 1970s, the National Aeronautics and Space Administration (NASA) created a robotics Mars Rover that could explore the foreign surface of another planet [33, 303, 305]. The first industrial robot was a hydraulic arm, purchased by General Motors in 1954 and able to lift heavy loads. In the same year, Barrett Electronics Corporation produced the first autonomous guided vehicle. Over time, locomotion, manipulation, and sensing abilities of robots continued to improve. In recent decades, the world has seen autonomous drones, vehicles capable of driving autonomously on standard roads, and humanoid robots capable of performing acrobatics (i.e., Boston Dynamics' Atlas [213]). Robots are used in factories, warehouses, delivery, homes, operating rooms, caregiving contexts, and for diverse applications in agriculture, healthcare, and manufacturing.

Crucial to more recent and ongoing advances in robotics is the field of artificial intelligence (AI). AI involves giving a computer or robot capabilities typically considered to require human intelligence, and always involves the machine engaging in behavior or outputting results that are not the result of explicit programming. Although much could be written on the intersection of robotics and artificial intelligence, in this monograph we focus on a subfield of AI known as reinforcement learning (RL).

Reinforcement learning, a type of machine learning, is growing in popularity and is being used to solve ever more complex problems [81]. We discuss RL in more detail in Sect. 1.2.1, but simply put, RL is a process by which a machine learns how to solve a task or accomplish a goal by experimenting with various ways of acting or achieving that goal, analyzing the results of those actions, and receiving feedback on its decisions. We call the machine an "agent," and if it is a robot, it is an "embodied agent." The agent takes action in an "environment," which can include both what the agent is capable of sensing as well as hidden information. The agent observes how the environment changes in response to its actions and receives a reward signal. (The reward signal is a property of the environment, often defined by a human.) The agent learns to solve the task by attempting to maximize the reward signal [290].

"Reinforcement" is the idea that as the agent acts, the reward signal it receives reinforces the desired behavior. (See Sect. 1.2.1 for more details.)

Reinforcement learning has been used in robots for terrestrial and aerial navigation, manipulation of objects, and autonomous driving in a variety of environments.

The apparent success of these techniques, however, at times outpaces our ability to understand why they make the decisions they do. Despite the increasing presence of autonomous robots and systems in everyday life, the internal working and reasoning of such systems can remain a "black box," in that we can see the inputs and outputs but not understand at a semantic level the internal processes that caused that output. Especially when using deep-learning learning algorithms (a type of learning technique that can be applied to reinforcement learning and other types of machine learning) [165], the agent will learn a knowledge representation or policy representation that it will then use to take actions. This representation is often, but not always, in the form of a neural network, sometimes with trillions of parameters. A machine can use this network to accomplish a task, sometimes discovering solutions of which no human has conceived (as in the case of AlphaGo's "move 37" [194, 283], where it played an especially creative move in a game of Go against the human world champion). The engineers and scientists who create such machines can tell you the algorithms used to create that representation, but the representation itself is often opaque. We see what the robot does, and perhaps it seems to be operating well, but we don't know why! This concept is relevant to RL and other types of artificial intelligence.

The field of Explainable Artificial Intelligence (XAI) aims to address this question of "why?" "Why did a machine take a particular action?" can be a difficult question to answer in the case of a learning machine. Machine learning techniques enable an agent to discern patterns to figure out how to solve a task based on observed or collected data without requiring an explicitly programmed or algorithmic solution ahead of time. The machine might develop or learn its own knowledge representation (description of the world) or policy (behavior plan). Sometimes, these learned representations cannot be understood by a human, even when humans understand the algorithms that create them and can test what inputs cause

1.1 Motivation

what outputs. When a system is trained using end-to-end machine learning, black box testing methods cannot validate the model with the same level of confidence as methods available when there is an explainable model [275]. Indeed, past work surveying and discussing multiple real-world applications of explanations for black box models has suggested that if a model for a high impact decision does not reveal its inner workings or provide an explanation, it should be avoided [253, 255]. This survey considers the intersection of explainable robotics and explainable reinforcement learning.

"Explainability" refers to the degree to which an AI system or method provides human-understandable reasons for its behavior or demonstrates what mechanisms underlie output generation, and "interpretability" means a human is able to inspect an agent's policy, understand it, and/or apply its outputs to a specific use case [43, 107].

An "explanation" conveys the procedure or means by which an output was generated, while an "interpretation" gives context that allows a human to understand its meaning [43].

Explanations can enhance trust [136], increase adoption [322], reduce risk [137], increase safety [69], and help with debugging [73]. Recently we have begun to see regulations such as Europe's GDPR [316], which require certain systems to have explanations, adding further urgency to our drive towards explainability.

DARPA has made explainable AI a priority in terms of allocating research funding and evaluating project outcomes [108].

Dulac-Arnold et al. [81] argues that among the challenges preventing real-world adoption of RL algorithms are the difficulties in preventing safety constraints from being violated and the problems faced by systems operators who desire explainable policies and actions. In addition, humans prefer to use systems that are trustworthy [245]. ("Trust" is related to but distinct from the idea of explanations. An explanation is a reason given for a decision. Trust is the concept that a human believes the explanation or believes that the system will make appropriate decisions even in the absence of an explanation. Explanations of some decisions can give a human confidence to trust unexplained decisions.) Explainability would also aid other types of RL such as transfer learning [355].

When the focus shifts from XAI in general to robotics applications in particular (explainable robotics), several unique considerations and opportunities arise. Robots are often physical systems. They can be used in isolated settings, as standalone systems in human settings, or to work jointly with humans. Robots may operate in proximity to humans regardless of whether the relationship is coexistence or collaboration. The multiple modalities of sensory input and interactions with humans offer risks (for example, failure could result in physical injury) and possibilities for explanation (for example, motion and speech could be used for communication). Issues of safety become important. Consumers of robots may demand guarantees that are possible with traditional control and planning methods but are more difficult or impossible to achieve with learning methods.

As robotics and artificial intelligence continue to advance, the importance of including explainability in these advancements grows.

Explainable AI (XAI) in general has been extensively investigated and studied [78, 87, 114, 233]. To a lesser degree, there have been separate surveys into XAI in the context of RL, or explainable reinforcement learning (XRL) [124, 200, 238, 326] and explainable robotics [20, 257, 258]. This survey focuses on the landscape of explainability and interpretability in reinforcement learning for robotics and introduces a unified classification system for discussing this interdisciplinary research area.

The classification system we propose will help facilitate discussion of and research into this interdisciplinary field.

We first discuss twelve *attributes* by which methods can be described. These attributes are properties of XRL-Robotics methods (or XRL, XAI, or X-Robotics methods), which are descriptive, and which enable a person to understand some high-level information about the method at a glance before looking into the details. The attributes are meant to be useful for a person looking for a specific type of method for practical use, or for those searching for similar or contrasting methods against which to compare. A newcomer to the filed can also review the attributes to get a sense of the breadth of methods that exist and an understanding of the important concepts in this field. Potentially, reviewing the attributes and methods they describe can spark ideas to fill gaps in the literature or new research questions. The decision process by which we developed the specific attributes and their elements is described in Chap. 2.

Our classification system also classifies the existing methods into different *categories*. When methods are conceptually similar, we group them together into a category, as described below and in Chap. 3. Chapter 3 goes into detail on the XRL-Robotics landscape in an organized manner.

This survey identifies and proposes the following twelve attributes, which can be used to classify existing methods within the area XRL for Robotics:

i. **Model-Agnostic or Model-Specific (MAMS)**–Can the method be applied to any model out-of-the-box (agnostic) or is it restricted to a certain kind of model (specific)?
ii. **Self-Explainable or Post-Hoc (SEPH)**–Does the method produce a model with explainable properties or is the method applied to a potentially (but not necessarily) black box policy?
iii. **Scope**–Does the explanation cover all inputs/outputs, the whole model, or a specific output?
iv. **When-Produced**–Is the explanation produced before, during, or after the learning takes place?
v. **Format**–What is the format of the explanation? (For example, text, image, numerical weights of features.)
vi. **Knowledge Limits**–Does the algorithm/model/system know what it can or cannot do, under what constraints, and when it may fail?
vii. **Explanation Accuracy**–Distinct from the accuracy of the system output, how accurate is the explanation regarding system output?

1.1 Motivation

viii. **Audience**–For whom is the explanation intended? (For example, end-users, domain-specific end-user, engineers, executives, regulatory agencies.)
 ix. **Predictability**–Can we predict what a robot will do ahead of time?
  x. **Legibility**–Does robot behavior (actions the robot takes) imply its goals?
 xi. **Readability**–Does robot behavior (actions the robot takes) imply its next actions?
xii. **Reactivity**–Does a robot react to the environment or plan ahead?

Attributes are further explained in Chap. 2, with an additional reference in Table 2.1. Methods can be classified individually, or a group of methods can be classified according to these attributes. These attributes are drawn from or inspired by work from within the Explainable AI, Explainable Reinforcement Learning, Explainable Robotics, and Robotics fields of research and are proposed or included if we determine them relevant to XRL for Robotics. Where consensus exists among existing literature regarding the definition of a term or the use of a particular term for a particular concept, our classification system incorporates terminology and divisions standard in the field. Where standard concepts exist with divergent terminology, our survey notes this. These attributes describe essential characteristics of methods that are useful for beginning to understand any of them. Collecting the methods in this survey according to these attributes also highlights areas that merit further investigation.

The attributes can be used to understand the uses and limitations of a method, as well as to compare one method to another. If a researcher or engineer is looking for a particular type of method, evaluating methods by these criteria or referring to this survey where methods' attributes are clearly noted may be helpful.

In addition to the above attribute classification system, we categorize methods in Chap. 3. These categories group similar methods for the ease of reading this survey (e.g., decision tree methods, saliency methods). Some categories have sub-categories (e.g., single tree vs. multi tree decision trees). Sometimes, methods in a single category will all share a certain kind of classification on one or more of the attributes (for example, single decision trees are all Self-Explainable). Many future methods yet to be designed will fall into the categories noted here, and thus insight about existing methods in these categories may be able to be applied to future methods in the same category.

Categories are intended to allow a researcher to more easily conceive of, understand, and talk about the range of existing methods in this field. Methods are grouped into a category when they have similar implementations or other technical features in common. The 12 attributes in this survey all measure an aspect of explainability or interpretability. Attributes are intended to allow a researcher to evaluate and determine the degree or kind of explainability of a method or group of methods on one or all of these axes.

The contributions of this survey include:

1. A review of how related terminology is used in the field and past proposed classification systems.
2. A novel classification system to classify prior work into 12 important attributes, identifying consensus where such exists and identifying the best existing terminology otherwise.
3. A broad survey of the state of the art in XRL for robotics, divided into categories and subcategories, with descriptions of some methods, and subcategories or categories described according to the attributes noted.
4. Identification and proposals of areas for future research, including specific suggestions.

In Sect. 1.2, we discuss background on reinforcement learning, explainable AI, and explainable robotics. In Sect. 1.3, we describe the criteria we use to determine whether a paper is included in the survey. In Chap. 2, we describe past attempts at classifications and attributes for XRL or X-Robotics and propose our own classification scheme for XRL-Robotics. In Chap. 3, we review and describe the state of the art of major categories of XRL for Robotics. In Chap. 4, we present a discussion of additional considerations. In Chap. 5, we outline proposals for future work and offer concluding thoughts.

1.2 Background

This survey addresses a topic at the the intersection of robotics, reinforcement learning, and the study of explainability within each of those fields. Brief relevant background reading for each are mentioned below.

1.2.1 Reinforcement Learning

Reinforcement learning involves an agent learning how to solve a task by interacting with a dynamic environment and receiving feedback via a reward signal the agent attempts to maximize [290].

In addition to impressive non-robotics accomplishments such as superhuman performance in Go [278] and Atari games [198], reinforcement learning has been applied to robotics applications such as self-driving cars [148, 259, 295], dexterous manipulation [19, 216], and other robotics applications [150, 235, 351].

1.2 Background

1.2.2 Reinforcement Learning Versus Supervised, Unsupervised, and Imitation Learning

Reinforcement Learning is a type of machine learning and is distinct from other types of learning such as supervised learning, unsupervised learning, and imitation learning.

Supervised Learning (SL) [64] involves training a model on an existing labeled dataset. In RL, the training data comes from interacting with an environment. An RL agent must sometimes begin taking actions before anything is learned. Supervised learning usually does not have a concept of "taking actions," unless it is used to train a model that involves actions; in contrast, RL is implicitly task- and action-oriented. Sometimes certain optimization techniques could be used in both SL and RL, especially in the training of off-policy RL [210], where data collected during environment interaction is stored and processed externally from the environment.

Even so, in RL, the off-policy learning algorithm's purpose is to develop a policy that can solve a task in an environment. In contrast to evaluating a dataset, an interactive process involves an environment or state that can change in response to actions. (Sometimes part of the state may be hidden, so that simply viewing the state is not enough to determine optimal action.) Additionally, testing the result of an action requires interacting with that environment and incurring the result as well as any side effects (such as environmental changes or termination of the experimental run). An RL framework is suitable for interactive processes and deals with challenges and dangers of having to interact with an environment to collect data and test a policy, challenges an SL framework is typically unequipped to handle. The classic SL tasks are classification and regression, although it has been used for many other purposes as well.

The goal of Unsupervised Learning [28] is to find patterns and learn information from data without specifying a specific objective ahead of time and/or without having a labeled dataset. For example, an unsupervised Natural Language Processing (NLP) clustering algorithm might determine what topics are discussed in various documents [27, 126].

Imitation Learning (IL) [86, 134, 164] requires a pre-existing model or policy. The goal of an IL algorithm is to imitate this model. Potentially, this would be transforming a model from one format to another or learning from an observation or demonstration.

Both UL and IL differ from RL in that RL typically has a specific objective to optimize. RL differs from IL and UL in many of the same ways it differs from SL, including the focus on environment interaction and tasks. IL might also involve solving a task, but if it does so then it does so through the idea of matching the performance of an oracle or expert instead of exploring the space of possible solutions and actions to discover a better, more optimal policy.

Supervised learning, unsupervised learning, imitation learning, and inverse reinforcement learning can all be used in conjunction with reinforcement learning as well, and several techniques discussed in Chap. 3 incorporate multiple forms of learning.

1.2.3 Explainable Artificial Intelligence (XAI) and Explainable Reinforcement Learning (XRL)

Beyond simple interest, opaque systems have negative consequences in the real-world. In healthcare, for example, it is difficult to deploy a system that doctors cannot trust [55]. Researchers focused on AI and financial securities have described scenarios in which an AI learns how to generate a profit. Due to the inability to scrutinize the strategy, it can be difficult to tell if it is engaging in market manipulation or not (several court cases of this nature have been brought, but convictions are rare) [29]. When we a deploy an AI in a machine with the ability to injure, such as an autonomous vehicle, failure to perceive what the AI has learned results in risks to human safety [220].

Transparency, which has several definitions but generally means a mechanism to gain insight into the inner workings of a policy or system, is important for generating trust in a system [103]. Transparency enables us to verify a model and ensure that it is fair, ethical, and working as intended [4].

Regulations such as Europe's GDPR [316] discuss the "right to an explanation." This makes explanation and interpretation important for algorithms and systems that developers intend to be used in the real world in Europe. The GDPR also discusses a right to "safeguards" such as human intervention. If an electronic action of an AI would potentially have legal effects or affect the data collected from a person, there must be a human in the loop (unless the subject has given informed consent).

There is often a tradeoff between interpretability and performance [78, 238] (also referred to as a "readability-performance tradeoff," although the word "readability" is used here to mean "interpretable" which is different from how the word is used in a robotics context, as described in Sect. 2.3.3).

Some methods involve attempting to use a white box instead of a black box model, while others seek to interpret the black box model [106].

Past surveys of XAI are numerous [78, 87, 114, 233]. Surveys of XRL exist as well [10, 102, 124, 154, 167, 195, 238, 298, 318, 344], although these do not focus on robotics as our survey does. Dulac-Arnold et al. [81] survey reinforcement learning in the real world in general. The National Institute for Standards and Technology (NIST) set forth four principles for XAI [233]. (For a discussion of the four principles and how they fit into our classification system see Sects. 2.1–2.3.)

1.2.4 Explainable Robotics (X-Robotics)

A user study has shown that autonomous robot systems with explanations can increase trust in that system and improve human-robot team performance [322]. Hepworth et al. [122] give a theoretical framework for enforcing transparency in human-robot swarm teaming. There have been surveys of explainable robotics [20, 257], but they do not focus on X-Robotics in the context of reinforcement learning, with the exception of [62] which is very brief.

1.3 Selection Criteria

We consider a paper relevant for our survey if it discusses (whether as its main focus or in a subsection) topic(s) that satisfy all of the following criteria:
1. Related to Reinforcement Learning,
2. Related to explainability or interpretability in the context of an RL algorithm or system in which RL plays a key component, and
3. Related to Robotics: the method in question has been either (i) applied to a robotic application (whether in simulation or physical world), or (ii) the method has not yet been applied to a robotic system, but the authors believe that it could be advantageously applied to a robotic system.

We consider papers that propose a new method including or providing interpretability or improving an existing method, as well as surveys and studies of such.

Classification System

2

In this chapter, we discuss the decision process behind selecting the 12 attributes we use in our classification system. These 12 attributes can be used to understand the ways in which a particular method is explainable and for what purposes it could or could not be used. The attributes provide a standard language for talking about explainability. As noted, they are drawn from the literature of both explainable artificial intelligence (machine learning, reinforcement learning) and robotics.

In Sect. 2.1, we discuss prior surveys that are relevant. In Sect. 2.2, we discuss the terminology used in prior work and note which of it we use or eschew in our work. In Sect. 2.3 we present a detailed description of the 12 attributes, how they are defined, and what they measure. A summary of the 12 attributes is given in Table 2.1.

2.1 Prior Surveys on XRL or X-Robotics

To the best of our knowledge, there are no previous surveys on Explainable or Interpretable Reinforcement Learning for Robotics. There are, however, surveys on Explainable Reinforcement Learning and surveys on Explainable Robotics.

Alharin et al. [10] presents a comprehensive survey of XRL. It covers many of the categories and subcategories we include in this survey (sometimes under different names), but it does not include Observation Analysis: A/B Testing (Sect. 3.5.3), Observation Analysis: Training Data (Sect. 3.5.4), Hierarchical Primitive Generation (Sect. 3.9.2), Causal Methods (Sect. 3.12), Reward Decomposition (Sect. 3.13), Symbolic Methods (Sect. 3.16), Instruction-Following (Sect. 3.15), or any robot-specific explanations such as Legibility (Sect. 3.17). Alharin et al. [10] also does not specifically call out visualizations (Sect. 3.14).

Alharin et al. [10] considers all five Hard Attributes (Sect. 2.3.1), and Explanation Accuracy (which they call Fidelity), but no other Attributes from the 12 we collect here.

Puiutta and Veith [238] is an XRL survey that identifies all the Hard Attributes aside from Explanation Format, and considers Predictability (which they call Interpretability), but this work does not discuss any other attributes from those we consider. Most of the categories of methods we cover are left unexamined by [238], which covers Soft Decision Trees, Linear Model U-Trees, Observation Analysis: Statistical or Frequency Techniques (which they call Interestingness, a term from one of the prominent papers in this subcategory), Programmatically Interpretable RL, Hierarchical Skill Acquisition, Causality, and Reward Decomposition. Puiutta and Veith [238] self-describes as non-exhaustive.

Heuillet et al. [124] is an XRL survey that does not consider any of our 12 attributes aside from intended Audience. The categories of methods [124] surveys comprise a minority of the categories we survey and include Saliency Maps, Representation Learning, Interaction Data (what we call Observation Based, Sect. 3.5.1), and Hierarchical Goals. They also discuss safe RL, as we do. Heuillet et al. [124] discusses Structured Causal Models (SCM), reward decomposition, and model uncertainty under the single heading "Simultaneous Learning," whereas in our work they are separate categories. Heuillet et al. [124] also discusses imitation learning and implies that imitation learning itself is explainable, a perspective with which we disagree, although explainable methods can certainly include imitation learning.

Wells and Bednarz [326] describes XRL methods in the areas of visualization, query-based explanation, policy summarization, verification, and human-in-the-loop collaboration. That work covers some of what our survey covers, although our survey is broader, and that survey does not address the twelve attributes we have identified. In addition to identify existing methods, [325] discusses some limitations of XRL method papers, such as the frequency of using toy examples instead of more complex experiments and lack of user studies. Methods in our survey suffer less from the first problem since most robotics experiments involve a certain amount of complexity. Lack of user studies is an issue in the Explainability field, although there are several relevant studies we highlight throughout this survey as relevant.

NIST published a document concerning Explainable Artificial Intelligence [233]. It is not a survey but rather identifies principles to keep in mind in XAI. Even though it is a broader scope that the other surveys we discuss in this section, we include it because it introduces the concept of "Knowledge Limits," which other surveys do not consider explicitly. Philips et al. [233] additionally discusses Explanation Accuracy, Audience ("Type of Explanation"), Self-Explainable vs. Post-Hoc, and Scope. The NIST document also includes a discussion of adversarial attacks on explainability (out of scope for our work) and discusses how a human can evaluate one's ability to explain machines by comparing it to one's also imperfect ability to explain humans.

Sado et al. [257] is a survey on Explainable Robotics. Naturally, [257] considers Predictability, Readability, and Legibility, as we do, unlike the pure XRL surveys. However, they do not consider any other of our 12 attributes explicitly aside from Reactivity. The only

2.2 Existing Classification Terminology for XRL or X-Robotics

overlapping categories with our survey are Observation Analysis and Causality (because the survey is not focused on RL, but robotics in general).

Anjomshoae et al. [20] surveys explainable techniques for "goal-driven agents" (i.e., robots). Anjomshoae et al. [20] is concerned with an agent communicating to a human such that the human can build a Theory-of-Mind [45] of the robot agent. They cover a wide range of methods, but do not cover any of our Attributes aside from Explanation Format, which they term Presentation.

Sakai and Nagai [258] is an X-Robotics survey that is less focused on individual methods and more on the general concept of what makes a useful explanation for a robot. They seek to formally define explainability, which we explicitly avoid in our work (our rationale is described in Sect. 2.2).

2.2 Existing Classification Terminology for XRL or X-Robotics

In this section we discuss different evaluation metrics and attributes and describe which ones we officially include and utilize in our survey. If we do not include a metric/attribute, we explain why in this section. If we do include an attribute in our survey we describe it in more detail in Sect. 2.3. In general, we seek i) to use language that seems to be dominant in the literature, where such exists, and ii) to eschew formally using language that seems to have widely differing meanings ("interpretability," "explainability," "transparency").

Most surveys of XAI or a subset of XAI describe the **Scope** of a method, which refers to whether explanations are global or per-decision [10, 124, 238]. We include Scope as a metric in our classification system and discuss it further in Sect. 2.3.

It is also common to see discussion of whether a method is self-explainable or post-hoc. This refers to whether a human can inherently understand a policy or whether it requires some additional process to produce the explanation. The former case is referred to as self-explainable [233] or intrinsic [10, 47, 238]. The latter is referred to as post-hoc [10, 124, 238] or simply "not self-explainable" [233]. Most surveys do not give this attribute a name, aside from [10], which calls it "type." We feel this is not descriptive and refer to it as the property of being "Self-Explainable or Post-Hoc." We discuss it in more detail in Sect. 2.3.

In a similar vein, the XAI literature discusses a method being model-agnostic or model-specific [10, 238]. We incorporate this term, using the acronym **Model-Agnostic or Model-Specific (MAMS)**.

More rarely, past work considers the explanation **Format** [10] (also referred to as the "presentation" [20]). Explanations can be numbers, text, visual augmentations, or new visualizations. Of note is that, in the context of robotics, additional formats are available such as expressive motions, light, and speech. (Anjomshoae et al. [20] considers log files to be a type of explanation as well, although we do not generally consider that case in our survey.) Format is an important attribute and we include this in our survey.

In Sect. 2.3 we will talk about an attribute called "**When-Produced**," which is our name for "when the explanation is produced." This is important to consider and has been discussed in past works but without consistent terminology. Alharin et al. [10] refers to "Order of Applying the Model" and gives three values for it: "before," "during," and "after." (We use the same terminology for categorical elements but have two "during" elements instead of one: "during-intrinsic" and "during-byproduct." See Sect. 2.3 for details.) [238] refers simply to "when the information is extracted" as an important criterion (in Sect. 2 of their paper where they describe their taxonomy), but it is not revisited or defined in that section or otherwise.

NIST introduces a concept called **Knowledge Limits**, which refers to a system's ability to know what it does and does not know or is capable of [233]. We use this concept in our work.

Distinct from the accuracy or performance of a model or system, one can discuss the accuracy of an explanation itself. This can be referred to as **Explanation Accuracy** [233] or Fidelity [10]. The latter asserts that this, the degree to which an interpretation is truthful to the actual model, is most important for post-hoc methods. We include Explanation Accuracy in our classification system.

It is also important to consider the **Audience** or the Purpose of the explanation. Heuillet et al. [124] and Murdoch et al. [211] discuss how the intended audience could be experts (medical doctors or insurance agents), users, developers, executives, regulatory agencies, or others. NIST and others discuss how different systems can satisfy various purposes such as user benefits, societal acceptance, regulatory compliance, or system development [35, 113, 233, 325]. The importance of considering the audience in XAI in general has been studied [23]. A system can ignore consideration of the audience, or it can be aimed at a specific audience. Another option is to customize explanations according to what is most helpful to an expert or non-expert, as [68] does by learning a "context of a failure and history of past actions." We note that this is an important attribute, but we do not attempt to classify methods based on it because it is too often subjective.

There are three robot-specific attributes that we note in this survey: **Predictability, Legibility, and Readability** [257]. These all relate to the robot communicating information to a human or matching expectations by virtue of the actions the robot takes. We discuss these attributes further in Sect. 2.3, since we include them in this study. An overview of some of these concepts can be found in [48]. Legibility, specifically, is discussed in [36, 80]. Readability is discussed in [88], although confusingly they use the term "predictability" to refer to what we describe as "readability." We use the terms "predictability" and "readability" according to what we found to be the more generally accepted definitions. We also note that predictability (as we use the term, and we use the definition from [257], described in Sect. 2.3) is similar to what [238] called "interpretability." We do not use the word "interpretability" or "explainability" for any one single attribute, as all attributes relate to these concepts, and these words are used in a wide variety of ways across the literature.

2.2 Existing Classification Terminology for XRL or X-Robotics

In the robotics literature, legibility, predictability, and readability regularly come up in contexts where no learning is involved. Randhavane et al. [242] is an example of a non-RL approach to making robots indicate their intent to humans by moving in a socially-aware way. In contrast, we focus on where these concepts intersect with reinforcement learning.

Another robot-specific attribute is **Reactivity**: whether a robot is reactive (it reacts to changes in its environment), deliberative (searches a space of behaviors and generates a plan), or some hybrid of the two [257]. We include this attribute.

Past work describes "Transparency" in a variety of ways. Sado et al. [257], Wortham [330] define it as the ability to describe, inspect, and reproduce the mechanisms through which an AI agent/robot makes decisions and learns to adapt to its environment. Puiutta and Veith [238] does not provide its own definition for transparency but notes that [178] uses transparency as a synonym for interpretability; [91] calls it comprehensibility. There are further definitions for transparency in [51, 76, 77, 91, 202]. We do not use the term "transparency" in this survey because its usage varies so widely and all the concepts it conveys are covered by the attributes that we do incorporate.

"Explainability" is another term with widely different definitions. To our thinking, many terms in this field fall under the concept of explainability, but some works use it to mean something more precise. The word is used as an official attribute by [226, 233, 238, 257], and [20] refers to "Explainable Agency." Sado et al. [257] further defines "explicability" [49] as subtly distinct from "explainability" (the latter referring to robots making decisions that humans understand and the former referring to making plans that humans understand). Gilpin et al. [101] attempts to distinguish between explainabilty and interpretability. We consider many of the concepts in this survey to be related to "explainability." Since it is such a vague term used in differing ways, it does not appear in our official classification system as an attribute. The concepts that other works use it to connote are covered by the more specific and precise terms and language we use. (We do use "explainability" and "interpretability" as more general, imprecise terms that can refer to the broad range of methods covered in this survey.)

Another subjective attribute we came across was whether an explanation was "Meaningful;" that is, whether it makes sense to a human [233]. Sado et al. [257] has a similar concept called "understandability." While clearly very important, it would be difficult to measure and so has been excluded from this survey.

Past work has also discussed the "level of detail" of an explanation and the "time required to process" [233]. These are often subjective (although not always) and have been excluded from the survey.

A "level of evaluation" is described in [124], originally from [77]. This measures whether the explanation is targeted at explaining a model (function level), a human layperson (human level), or an end user (application level). (Heuillet et al. [124] distinguishes between layperson and end user.) We do not include this attribute in our survey because it is imprecise and requires the development of a system. Some methods described in this survey could be used at any of the application, human, or function levels. In some cases, a method that applies to one level could be made to apply to another with some engineering. To the extent that this attribute could be formalized, it overlaps with Audience and Format, and we feel these latter two attributes better capture the attributes under discussion. Thus, we exclude "level of evaluation" as an attribute from this survey.

There are some additional robot-specific attributes that others use that we exclude from the survey. Anjomshoae et al. [20] discusses "application," that is, the end-user application such as search and rescue or e-health. Many of the methods we discuss in our survey could be applied to multiple applications, and so we do not feel this is a useful attribute to consider. Anjomshoae et al. [20] discusses "dynamics," by which they mean the degree to which a robot is aware of users and aware of context and takes those aspects of the situation into account. We feel this is also too tied to specific applications to be relevant.

2.3 The Attributes of our Classification System

One of the contributions of this survey is to refine and present a system of classifying the methods that presently exist. We do this by creating **Categories** of methods, which are presented in Sect. 3, and by creating or incorporating attributes, discussed in this section, by which methods or categories can be compared to other methods or categories. Based on reading existing surveys and the authors' own investigations, the authors developed **attributes of comparison**, which are important attributes by which methods can be grouped or important considerations for evaluating an explainable robotics system. These are shown in Table 2.1.

In Table 2.1, we divided the attributes into three groups. **Hard attributes** are attributes that can generally be objectively defined, such as the *explanation format* (i.e., we know if the output is numerical, textual, or visual). There are two groups of **Soft attributes**, one for those applicable to XRL in general, and one group of **robot-specific attributes**. Soft attributes are attributes that are imprecisely defined, but which multiple humans might come to a similar consensus regarding. For example, the *intended audience* for the explanations from an AI system is not well-defined, but humans might have a similar, imprecisely bounded perception of what that audience could or should be. To give a robot-specific example, it is tough to precisely measure a robot's *readability*, that is, its ability to indicate to humans its future actions by virtue of its behavior. Despite being difficult to measure (although a user study can help), a given human interacting with a given robot might have a consistent sense of the degree to which this attribute was achieved. Not included in the table or our evaluation is what we call **Subjective** attributes. The subjective estimates of these attributes may vary widely even between humans. An example of a subjective attribute is how *meaningful* a robot's explanation is.

2.3 The Attributes of our Classification System

Table 2.1 The 12 Attributes of Comparison

Hard attributes	Model-Agnostic or Model-Specific	Is the method **model-agnostic** or **model-specific**?
	Self-Explainable or Post-Hoc	Does the method produce a policy or other artifact that is **intrinsically self-explainable** (can be directly interpreted), or is it a process that is applied to a policy **post-hoc**?
	Scope	Is the explanation **global** (full policy) or **local** (per-decision)?
	When-Produced	Where and how is the explanation produced? Is it a transform or other calculation applied **after** training? Is it an **intrinsic** attribute of the policy such that it is learned **during** training? Is it produced **during** training as a **byproduct** of the learning procedure but not used directly by the agent for decision-making? Or is it a result of preprocessing applied **before** training the model?
	Format	What is the format of the explanation (i.e., text, image, rules, lists, numerical)?
Soft Attributes (general)	Knowledge Limits	Does the system understand where it is applicable and designed to operate, and where it is not and inform the user? Does it provide a notice that its output may be inaccurate? Where applicable, we can describe satisfying this attribute as **not attempted, partially attempted**, or **completely attempted**.
	Explanation Accuracy	How much uncertainty do the explanations have? Where applicable, they could be **certain, uncertain (measured)**, or **uncertain (unmeasured)**
	Audience	What is the intended audience for the explanation i.e., end-users, domain-specific end-users, engineers, executives, regulatory agencies)?
Soft Attributes (robot specific)	Predictability	For a robotics system, can we predict what the robot will do, either by analyzing its policy or by watching it in real time? Where applicable, we can describe satisfying this attribute as **not attempted** or **attempted**
	Legibility	For a robotics system, does robot behavior imply its *goals* to the user or nearby humans? Where applicable, we can describe satisfying this attribute as **not attempted** or **attempted**
	Readability	For a robotics system, does robot behavior imply its *next action* to the user or nearby humans? Where applicable, we can describe satisfying this attribute as **not attempted** or **attempted**
	Reactivity	For a robotics system, where applicable, does the robot **react** to an environment or plan in a **deliberative** manner? It could also take a **hybrid** approach

2.3.1 Hard Attributes

Model-Agnostic or Model-Specific (MAMS) is a measure indicating whether an explainable method is model-specific or model-agnostic. For example, observation-based explanations can be applied to any underlying method and so are model-agnostic. Reward decomposition, which looks at weights of a specific reward function, are tied to that specific model and so are model-specific. Naturally, a model-specific method might be widely applicable, but additional effort must be taken for each new case (i.e., decomposing the reward function), whereas a model-agnostic method can be applied across different methods and environments with less customization. MAMS is an attribute that is commonly referenced in works on interpretability with consistent terminology for the categorical elements of this attribute (agnostic vs. specific). In addition to agnostic vs. specific, we note that the MAMS can be variable (not always exclusively one or the other).

Self-Explainable or Post-Hoc (SEPH) is a measure of whether a method or system is inherently explainable or whether explanations must be derived post-hoc [10, 47]. For example, a decision tree is intrinsically self-explainable, in that the very structure that exists and is used to map from features to output clearly shows why that output was reached. In contrast, a saliency map created via input perturbation (see Sect. 3.2.3) is an example of a post-hoc explanation, since it is an explanation that is constructed by a secondary process external to the learning method and learned policy themselves. Most previous surveys on explainability recognize this self-explainable/post-hoc distinction (sometimes referring to "self-explainable" as "intrinsic" or "inherent").

Scope is a measure of how much of the policy the explanation explains. We say that an explanation is global or local. Global explanations attempt to elucidate the entire general logic of behavior of a model or policy, while local explanations provide insight into a specific decision or group of decisions [4, 201]. To re-use the previous example, a decision tree (where the explanation is the policy) is a global explanation, meaning that it explains the entire policy, whereas a saliency map (typically) explains one particular state-action pair (one local decision in time). It does not match up directly to SEPH, however. Certain uses of Representation Learning (see Sect. 3.4.2) can produce self-explainable explanations that are local, and analyzing interactions/observation-based (see Sect. 3.5) methods for explanation can produce global explanations in a post-hoc manner. We also consider cases where scope can be both local or global and cases where it could be used for either type of explanation.

When-Produced is a measure of *when* the explanation is generated. This is a less-commonly used attribute in the literature and there is no consistent terminology for it (see Sect. 2.2 for a comparison). We propose the following definitions adapted from [10]:

1. "Before": the interpretation is applied or produced before training the model.
2. "During-Intrinsic": the explanation is produced via the process that creates the policy, and the explanation is a necessary component of the policy used for execution.
3. "During-Byproduct": the explanation is produced via the process that creates the policy, but the explanation is not necessary for execution, making it a separate product from the policy.
4. "After": a post-training transform, calculation, or other process is used to create the explanation.

2.3 The Attributes of our Classification System

Sometimes a method may have elements of more than one When-Produced type.

An example of When-Produced:Before is the Instruction Following category of methods (see Sect. 3.15) since the instructions later used to construct explanations precede the training of the policy and the production of explanations.

An example of a When-Produced:During-Intrinsic method would be one that directly learns a decision tree or symbolic model. A learned symbolic model (see Sect. 3.16.3) represents the policy using symbolic rules and/or a high-level abstract graph that a human can follow that a robot also uses for its own execution. When-Produced:During-Byproduct is similar to When-Produced:During-Intrinsic with one crucial distinction: the explanation is not necessary for execution. There is a method that determines (during training) important trajectories of which to record a video to be visualized and provide understanding of the training process to the user (see Sect. 3.5.1). These explanations are produced during training and are ready when the policy itself is ready, but they are necessary only for human understanding and not required for learning or executing the policy itself. Another example is the method for learning counterfactual explanations during training, discussed in Sect. 3.3.1.

An example of the When-Produced:After would be the State Transformation technique discussed in Sect. 3.4.1, which results in an abstract Markov Decision Process (MDP) for human consumption. Model Reconciliation (Sect. 3.11) is also When-Produced:After.

Note that a method that is of the When-Produced type "During-Intrinsic" might not be Self-Explainable. Sometimes it may be Self-Explainable and When-Produced:During-Intrinsic, such as in the case of Conservative Q-Improvement [252], which uses RL to learn a decision tree (see Sect. 3.1.1). A different decision-tree-based method such as a policy distillation method like Multiple Scenario Verifiable Reinforcement Learning via Policy Extraction (MSVIPER) [250] (also in Sect. 3.1.1), however, would be Self-Explainable (since the end product is a decision tree representing the entire policy) but When-Produced type "After" (since policy distillation is used to transform expert policy into a tree). Additionally, the two examples given above for When-Produced:After are of Post-Hoc SEPH. While nothing precludes a method that involves Post-Hoc explanations from being When-Produced:During-Intrinsic, in practice we found that all methods producing explanations by means of a When-Produced:During-Intrinsic means were Self-Explainable, and that other types of When-Produced could produce either kind of Self-Explainable or Post-Hoc.

Explanation Format is the final Hard Attribute. Explanations come in a variety of different formats, so much so that we do not attempt to create a finite, categorical list. Explanations can be numerical, numerical weights, visual, or text of various kinds (labels, sentences from templates, or even more full language explanations). Some methods produce representative states or trajectories. Other methods produce policies with explanatory formats, such as a tree, logical procedure, or graph. Meaningful Representation Learning (see Sect. 3.4.2) transforms a state space into a form that a human can more easily comprehend, whether that be turning many dimensions into a few, or creating a graph. Saliency methods (see Sect. 3.2) augment a visual space with visual markers to give information to a human. All these formats are precisely defined and are critical components for classifying methods. A developer looking for an explanatory technique might want one that has a particular format. For example, a non-visual state space precludes saliency maps, and a system for a lay person might suggest using a textual explanation that might be unnecessary for a fellow engineer.

Each method and category in Sect. 3 is analyzed according to these hard attributes.

2.3.2 Soft Attributes (General)

Knowledge Limits is a concept introduced by NIST [233] and refers to the idea that a robot should know the limits of its own capabilities. Part of engendering trust and aiding proper usage of a system is managing expectations, and awareness of knowledge limits facilitates setting these expectations [151]. We extend this idea further and, in our work, classify methods according to this attribute across three types of Knowledge Limits: (i) Not Attempted, if the method makes no attempt to understand its own limits; (ii) Partially Attempted, if there is a partial communication of limits; and (iii) Completely Attempted, in which a method enables a robot to attempt to achieve full understanding of its limits. In practice, it is rare for a system to attempt to satisfy the idea of Knowledge Limits, but we do discuss some cases towards this end.

Explanation Accuracy refers not to the accuracy of a policy but to the accuracy of the explanation. Is the content of the explanation accurate; that is, does it reflect the underlying robot process? We assign three elements to this attribute: (i) Certain (completely accurate explanations), (ii) Uncertain-Measured (explanations may have inaccuracies but the method attempts to account for this and provide information on explanation accuracy), and (iii) Uncertain-Unmeasured (explanations may or may not be accurate and no attempt is made to communicate the degree of accuracy to the user). For example, a decision tree has Certain Accuracy, because the robot is using the tree itself for execution. In contrast, a category of methods involving using a template to redefine MDP-based RL policies in human readable language (see Sect. 3.10.1) gets assigned Uncertain-Measured because the explanations may not be precisely accurate but the method attempts to give an estimate of how certain it is in its explanations. Most methods surveyed have accuracy Uncertain-Unmeasured. For example, many saliency map methods give explanations for a black box policy using black box explanations.

One note on how we measure Explanation Accuracy in this survey: we evaluate based on our assessment of real-world accuracy. So, if a method would have theoretical certainty if given infinite samples but uncertainty if given finite samples (like Goal Driven Memory XRL in Sect. 3.13.1), we say that it is imprecise/uncertain. If this uncertainty is not quantified, we say it is unmeasured.

Audience refers to the intended audience for the explanation. For example, some systems are aimed at end-users, while others are meant to give information to a fellow machine-learning practitioner. We do not track this attribute across most of the methods in this survey, however, since it is not applicable to many methods and is sometimes subjective. We include it in our list because despite the subjectivity, it is important to consider in practice.

2.3.3 Soft Attributes (Robot-Specific)

The previous subsection described soft attributes that could easily describe any XAI method. Here we describe soft attributes that are specific to a robotics context.

Predictability refers to the idea that a robot's behavior or actions match what a human might expect it to do. (Note that this is NOT the same as a method that allows a human to predict what it will do from looking at the policy. Predictability in robotics literature refers to robot behavior at execution time.)

Legibility refers to the robot being intent-expressive, in that the actions of the robot communicate its intended future goals to an observer. More formal definitions of legibility can be found in [36, 80].

Readability refers to the robot being action-expressive, in that the actions of the robot communicate its intended future actions to an observer.

Reactivity is an attribute referring to whether a method creates a means for responding to the environment in the moment or formulates a plan and executes it. If it reacts, we call it "reactive," and if it plans, we call it "deliberative." Naturally, some take a "hybrid" approach.

Predictability, Legibility, and Readability are standard concepts from Robotics that allow human-robot interaction. They are understudied in the literature focused on XAI and XRL. It is of particular relevance for robotics applications.

Explainable Methods Organized by Category 3

In this section we describe specific state-of-the-art methods for XRL for robotics, organized by category.

Within each category and subcategory, we classify the category or subcategory according to the following attributes: scope, when-produced, explanation format, whether self-explainable or post-hoc, and whether model-agnostic or model-specific. (Unless otherwise noted, assume that knowledge limits are "Not Attempted" and the accuracy is "Uncertain-Unmeasured." We only note the reactivity if it is nonobvious (most RL policies are reactive; if planning is involved, we mention it). If we say nothing about predictability, legibility, or readability, assume that the methods in that category do not attempt to address those issues.)

Tables 3.1 and 3.2 describe the SEPH, Scope, MAMS, and When-Produced for certain methods. Tables 3.3 and 3.4 describe the Explanation Format, Knowledge Limits, and Explanation Accuracy for certain methods. In both tables, methods are grouped and cited, and sections (corresponding to categories/subcategories) are noted in which more detailed discussion can be found. One method might belong to multiple categories.

3.1 Decision Tree

One of the basic ways to provide an explanation is by using an explainable format. One classic explainable format is a decision tree [112]. Decision trees are easy to parse (when they are small enough) and can be analyzed by analytical methods [37, 71, 112]. In this section we discuss ways that decision trees are used to provide explainability to a robot reinforcement learning agent.

Table 3.1 Highlighted methods, grouped, and their attributes in our classification system (Part 1a)

Method name		Category/Sections	SEPH	Scope	MAMS	When - Produced
Using RL to learn DT by Additive Process	[192, 248, 252] [65, 306, 353]	3.1.1	SE	Global	MS	Dur-Int
Learn DT by Policy Distillation	[179, 249, 250] [75, 197]	3.1.1	SE	Global	MS	After
Learn Mixture of DT by Policy Distillation	[44, 309] [196]	3.1.3, 3.1.2	SE	Global	MS	After
Differential Decision Tree Policy	[225, 277]	3.1.1	SE	Global	MS	Dur-Int
Causal DT Explanation	[117, 189]	3.1.1, 3.12	SE	Global	MS	After
Learn Soft Decision Tree by Policy Distillation	[58]	3.1.2	SE	Global	MS	After
Saliency Visualization as Attention Side Effect	[219, 284, 340] [26, 204, 205] [239]	3.2.2, 3.14	SE	Local	MS	Dur-By
Saliency Visualization by Forward Propagation or Backpropagation	[186, 323] [133, 331]	3.2.1, 3.2.2	PH	Both	MS	After
Saliency Visualization by Input Perturbation	[89, 104, 336] [92, 111, 132] [66, 218]	3.2.3, 3.14	PH	Local	MA	After
State Transformation: Abstract MDP	[199, 341, 356] [25, 302]	3.4.1, 3.9.1, 3.14	PH	Global	MA	After
State Transformation: Clustering States for Abstraction	[8, 9]	3.4.1	SE	Global	MA	Dur-Int
State Transformation: Planning over Abstract Graph	[177]	3.4.2	SE	Global	MA	Dur-Int
Custom Domain Specific Language as Action Components	[312, 313]	3.6	SE	Global	MS	Before
Instruction Following	[156, 268, 310] [142, 145, 193] [57, 93, 94]	3.15	SE	Local	MS	Before
Human-in-the-Loop Correction	[183]	3.15	PH	Local	MS	Before
Symbolic Policy By Distillation	[128, 162]	3.16.1	PH	Global	MA	After
Symbolic Rewards by Evolution	[269]	3.16.2	SE	Global	MA	Dur-Int
Symbolic Model or Policy	[96, 279, 334]	3.16.3, 3.9.1	SE	Global	MS	Dur-Int
Improving Legibility	[36, 232, 352]	3.17	SE	Global	MA	Dur-Int
RL Visualization for Debugging	[73, 197]	3.14	PH	Local	MA	After

Each row is a method or grouping of similar methods discussed in Chapter 3: Explainable Methods by Category. The name is our description for it. The following attributes are noted in this table (and defined in Sect. 2.3):
· **SEPH**: Self-Explainable (SE), Post-Hoc (PH)
· **Scope**: Global, Local, Both/Either
· **MAMS**: Model-Agnostic (MA), Model-Specific (MS), Variable (V)
· **When-Produced**: Before, During-Intrinsic (Dur-Int), During-Byproduct (Dur-By), After

3.1 Decision Tree

Table 3.2 Highlighted methods, grouped, and their attributes in our classification system (Part 1b)

Method name	References	Category/ Sections	SEPH	Scope	MAMS	When-Produced
Safety-Constrained Learning	[3, 16, 345] [149, 299, 335] [59]	3.7	SE	Global	MS	Before
Safety-Informed Execution: Uncertainty Aware Action-Selection	[155, 187] [227, 294]	3.8	PH	Local	MS	Dur-By
Safety in Execution: Safety via planning	[82]	3.8, 3.4.2	SE	Global	MS	Dur-Int
High Level Interpretability via Hierarchical RL	[34, 191, 273] [143, 166, 279]	3.9.1	SE	Global	MS	Dur-Int
Primitive or Skill Generation for Hierarchical RL	[127, 142] [173, 332, 333]	3.9.2	SE	Local	MS	Dur-Int
Certain Model Reconciliation	[152, 153]	3.11.1	PH	Local	MS	After
Uncertain Model Reconciliation	[50, 285]	3.11.2	PH	Local	MS	After
Meaningful Representation Learning: Metadata or Human/Domain Data	[40, 41, 282]	3.4.2	SE	Local	V	Dur-Int
Meaningful Representation Learning: Learning Logical Relationships	[343]	3.4.2, 3.16.3	SE	Global	MS	Dur-By
Meaningful Representation Learning: Autonomously Discovered Latent States	[110, 203, 241] [90, 243, 348] [140, 212, 350]	3.4.2	SE	Global	V	Dur-Int
Meaningful Representation Learning: Graph Representations of Visual Input	[175, 274]	3.4.2, 3.16.3	SE	Either	MS	Dur-Int
Counterfactual Explanations (GAN)	[222, 223]	3.3.1	PH	Local	MS	After
Contrastive Explanations Learned During Training	[176]	3.3.1	PH	Local	MS	Dur-By

(continued)

Table 3.2 (continued)

Method name	References	Category/Sections	SEPH	Scope	MAMS	When-Produced
Counterfactual Explanations (SCM)	[189, 234]	3.3.2, 3.12	PH	Local	MS	Dur-By
Causal Influence Models excluding counterfactuals)	[85, 264]	3.12	SE	Both	MS	Dur-In
Counterfactual Explanations (non causal)	[54, 197] [95, 206]	3.3.2	PH	Local	MS	After
Contrastive Explanations as Justification	[286]	3.3.2	PH	Local	MS	After
Analyze Interaction for Policy Understanding	[2, 265, 266] [11, 14, 130] [133]	3.5.1, 3.14	PH	Global	MS	Dur-By
Analyze Interaction for Goal Understanding	[160, 161] [129, 131]	3.5.2	PH	Global	MS	After
Analyze Training Interaction for Transition Model	[69]	3.5.4	PH	Local	MA	After
Observation and Network Analysis: SHAP Feature Attribution for RL	[118, 174] [141, 184, 246]	3.5.4	PH	Local	MS	After
Interrogative Observation Analysis	[354]	3.5.5	PH	Local	MS	After
Analyze Interaction with A/B Testing	[267]	3.5.3	PH	Local	V	After
NLP Template for Model or Policy	[190, 322] [119]	3.10.1	PH	Local	MS	After
NLP Templates for Queries	[116]	3.10.2	SE	Local	MS	After
Reward Augmentation and Repair	[292]	3.13.2	PH	Local	MS	Dur-By
Minimal Sufficient Explanation w/Reward Decomposition)	[144]	3.13.1	PH	Local	MS	Dur-By
Memory-Based Probability-of-Success XRL	[61, 62]	3.13.2	PH	Local	MS	After

Each row is a method or grouping of similar methods discussed in Chapter 3: Explainable Methods by Category. The name is our description for it. The following attributes are noted in this table (and defined in Sect. 2.3):
· **SEPH**: Self-Explainable (SE), Post-Hoc (PH)
· **Scope**: Global, Local
· **MAMS**: Model-Agnostic (MA), Model-Specific (MS), Variable (V)
· **When-Produced**: Before, During-Intrinsic (Dur-Int), During-Byproduct (Dur-By), After

3.1 Decision Tree

Table 3.3 Highlighted methods, grouped, and their attributes in our classification system (Part 2a)

Method name	References	Category/Sections	Explanation format	KL	EA
Using RL to learn DT by Additive Process	[192, 248, 252] [65, 306, 353]	3.1.1	Tree	CA	C
Learn DT by Policy Distillation	[179, 249, 250] [75, 197]	3.1.1	Tree	CA	C
Learn Mixture of DT by Policy Distillation	[44, 309] [196]	3.1.3, 3.1.2	Mixture of Trees	CA	C
Differential Decision Tree Policy	[225, 277]	3.1.1	Tree	CA	C
Causal DT Explanation	[117, 189]	3.1.1, 3.12	Tree, Diagram, Text	N	C
Learn Soft Decision Tree by Policy Distillation	[58]	3.1.2	Tree	N	UU
Saliency Visualization as Attention Side Effect	[219, 284, 340] [26, 204, 205] [239]	3.2.2, 3.14	Pixels (individual features)	N	UU
Saliency Visualization by Forward Propagation or Backpropagation	[186, 323] [133, 331]	3.2.1, 3.2.2	Pixels (individual features)	N	UU
Saliency Visualization by Input Perturbation	[89, 104, 336] [92, 111, 132] [66, 218]	3.2.3, 3.14	Pixels (individual features)	N	UU
State Transformation: Abstract MDP	[199, 341, 356] [25, 302]	3.4.1, 3.9.1, 3.14	Simplified MDP Visualization	N	UU
State Transformation: Clustering States for Abstraction	[8, 9]	3.4.1	Simplified/Abstract States	N	UU
State Transformation: Planning over Abstract Graph	[177]	3.4.2	Graph of space	N	UU
Custom Domain Specific Language as Action Components	[312, 313]	3.6	Language	N	CA
Instruction Following	[156, 268, 310] [142, 145, 193] [57, 93, 94]	3.15	Instructions	N	UU
Human-in-the-Loop Correction	[183]	3.15	instructions	PA	UU
Symbolic Policy By Distillation	[128, 162]	3.16.1	Symbolic Structure	N	C
Symbolic Rewards by Evolution	[269]	3.16.2	Symbolic Rewards	N	UU
Symbolic Model or Policy	[96, 279, 334]	3.16.3, 3.9.1	Symbolic Structure	N	UU
Improving Legibility	[36, 232, 352]	3.17	Chosen actions are intent-expressive	N	n/a
RL Visualization for Debugging	[73, 197]	3.14	Visualization	N	UU

Each row is a method or grouping of similar methods discussed in Chapter 3: Explainable Methods by Category. The name is our description for it. The following attributes are noted in this table (and defined in Sect. 2.3):

· **Format**: (described per method)
· **Knowledge Limits (KL)**: Not Attempted (N), Partially Attempted (PA), Completely Attempted (CA)
· **Explanation Accuracy (EA)**: Certain (C), Uncertain (measured) (UM), Uncertain (unmeasured) (UU)

Table 3.4 Highlighted methods, grouped, and their attributes in our classification system (Part 2b)

Method name	References	Category/Sections	Explanation format	KL	EA
Safety-Constrained Learning	[3, 16, 345] [149, 299, 335] [59]	3.7	Rules or other constraints	N	UM
Safety-Informed Execution: Uncertainty Aware Action-Selection	[155, 187] [227, 294]	3.8	model uncertainty	PA	UU
Safety in Execution: Safety via planning	[82]	3.8, 3.4.2	planable latent space	N	UU
High Level Interpretability via Hierarchical RL	[34, 191, 273] [143, 166, 279]	3.9.1	Task plan	N	C
Primitive or Skill Generation for Hierarchical RL	[127, 142] [173, 332, 333]	3.9.2	skill choice	N	C
Certain Model Reconciliation	[152, 153]	3.11.1	Text explanations	CA	C
Uncertain Model Reconciliation	[50, 285]	3.11.2	Text explanations	N	C
Meaningful Representation Learning: Metadata or Human/Domain Data	[40, 41, 282]	3.4.2	Representation	N	UU
Meaningful Representation Learning: Learning Logical Relationships	[343]	3.4.2, 3.16.3	Representation	N	UU
Meaningful Representation Learning: Autonomously Discovered Latent States	[110, 203, 241] [90, 243, 348] [140, 212, 350]	3.4.2	Representation	N	UU
Meaningful Representation Learning: Graph Representations of Visual Input	[175, 274]	3.4.2, 3.16.3	Graph representation and/or symbolic plan upon the graph	N	C
Counterfactual Explanations (GAN)	[222, 223]	3.3.1	state (images) examples	N	UU
Contrastive Explanations Learned During Training	[176]	3.3.1	State examples with text explanations from manual features	N	UU
Counterfactual Explanations (SCM)	[189, 234]	3.3.2, 3.12	Text counterfacturals	N	C

(continued)

Table 3.4 (continued)

Method name	References	Category/Sections	Explanation format	KL	EA
Causal Influence Models excluding counterfactuals)	[85, 264]	3.12	Causal influence model	N	UU
Counterfactual Explanations (non causal)	[54, 197] [95, 206]	3.3.2	State examples	N	C
Contrastive Explanations as Justification	[286]	3.3.2	State examples, text explanations	N	UU
Analyze Interaction for Policy Understanding	[2, 265, 266] [11, 14, 130] [133]	3.5.1, 3.14	Video/trajectories of state action pairs)	N	UU
Analyze Interaction for Goal Understanding	[160, 161] [129, 131]	3.5.2	Trajectories	N	UU
Analyze Training Interaction for Transition Model	[69]	3.5.4	Feature weights	N	UU
Observation and Network Analysis: SHAP Feature Attribution for RL	[118, 174] [141, 184, 246]	3.5.4	Feature weights	N	C
Interrogative Observation Analysis	[354]	3.5.5	Hidden policy	N	UU
Analyze Interaction with A/B Testing	[267]	3.5.3	Hidden policy	PA	UU
NLP Template for Model or Policy	[190, 322] [119]	3.10.1	Text	CA	UM
NLP Templates for Queries	[116]	3.10.2	Text Respsone to Query	N	UU
Reward Augmentation and Repair	[292]	3.13.2	Number, transformed into words	N	UU
Minimal Sufficient Explanation w/Reward Decomposition)	[144]	3.13.1	Numerical weights of reward components)	N	C
Memory-Based Probability-of-Success XRL	[61, 62]	3.13.2	Number	N	UU

Each row is a method or grouping of similar methods discussed in Chapter 3: Explainable Methods by Category. The name is our description for it. The following attributes are noted in this table (and defined in Sect. 2.3):

· **Format**: (described per method)

· **Knowledge Limits (KL)**: Not Attempted (N), Partially Attempted (PA), Completely Attempted (CA)

· **Explanation Accuracy (EA)**: Certain (C), Uncertain (measured) (UM), Uncertain (unmeasured) (UU)

3.1.1 Single Decision Tree

> Self-Explainable; Global; Model-Specific; When-Produced: During-Intrinsic or When-Produced:After; Explanation Type: Tree; Has been applied to robotics in simulation and real world; Knowledge Limits: Complete; Explanation Accuracy: Certain

One way that a decision tree can be used is to represent the policy. The goal of reinforcement learning is to create a policy that maps between states and actions, and a single decision tree can represent such a mapping. In practice, it is more difficult to create such a mapping using a decision tree compared to a black box neural net because a neural net is typically a more powerful learner. (A neural net is more powerful in the sense that for high-complexity patterns or mappings, a neural net takes significantly less time to learn the solution.) A neural net can also often more efficiently represent a policy than a decision tree, if a decision tree could solve the given task at all. Nevertheless, several methods have been proposed.

First, we look at a few methods that use reinforcement learning to learn a decision tree by an additive process. In other words, they each start with the root node and then continue to add nodes node-by-node during the exploration process. Conservative Q-Improvement [248, 252] starts with a single root node and keeps track of hypothetical children nodes to create based on potential splits (a split being the breaking of a leaf node into a branch node and two children). At each time step it measures the estimated discounted future reward of the overall policy based on the current tree and the future hypothetical splits. If any split would yield an increase above a certain threshold, the split is performed. This approach yields a smaller tree policy with performance comparable to previous additive methods. (UTree [192, 306] and derivatives [353] are also additive approaches, but they require discrete state spaces.) Reinforcement learning is combined with genetic/evolutionary algorithms in [65] to build a tree and then adjust it genetically.

A differential decision tree policy [277] updates the entire tree during the learning phase instead of adding nodes incrementally as in previously discussed methods. They use a tree structure where gradient descent can be applied. The main difference between a differential decision tree and the normal decision tree is that the method in [277] uses a sigmoid function for splitting the nodes instead of a non-differentiable if/else rule or boolean condition. The leaf nodes in the resulting tree represent single features because the authors of [277] applied a discretization technique by replacing the feature vectors with the feature of maximum value. These ideas are extended and applied to simulated robotics environments by [225].

Another way to create a decision tree is to first create a black box policy as a neural net and then transform it into a decision tree using policy distillation. This is the approach taken by MSVIPER [249, 250], where a policy is first learned by standard reinforcement learning techniques. The resulting policy is then run in simulation and trajectories (sequences of state action pairs) are generated. These pairs form a dataset that can be used to create

3.1 Decision Tree

decision trees using standard techniques from supervised learning such as Classification and Regression Tree (CART) [169]. The state elements are the features, and the actions are the values of the leaf nodes. A similar transformation is applied by [179] for U-trees (and an application of linear model trees for autonomous maritime docking is found in [184]). [196] performs a multi-agent form of this tree distillation. [74, 75] use nonlinear decision trees (NLDT) to learn control roles for a 2D robotic arm simulation, and [100] uses an NLDT for a simulated lane-changing task. An NLDT [74, 75] is a tree of constrained size with a hierarchical "bilevel" optimization scheme. Splits are determined as a function of features, and parameters are fine-tuned on a lower level. This is a differently-optimized type of decision tree that seeks to match an oracle or expert policy, as do the other papers mentioned in this paragraph.

In particular, [249, 250] demonstrate one of the benefits of the decision tree policy, which is that the interpretability can be leveraged to improve the policy even beyond that of an expert policy, thus taking advantage of the learning power of a neural net and the interpretability of a decision tree. A diagram of this process is shown in Fig. 3.1. In [249, 250], these techniques are used to leverage interpretability to address robot freezing issues, reduce oscillation, and reduce vibration for mobile robots navigating among dynamic obstacles and outdoor robot navigation on complex, uneven terrains.

Another single-tree generation technique is found in [197], which uses a decision tree to create counterfactual explanations and visualizations. This technique is discussed further in Sect. 3.3.2. The Causal methods [117, 189] use a single decision tree to represent causality, as do many others, and are discussed further in Sect. 3.12. The observation-based method [2]

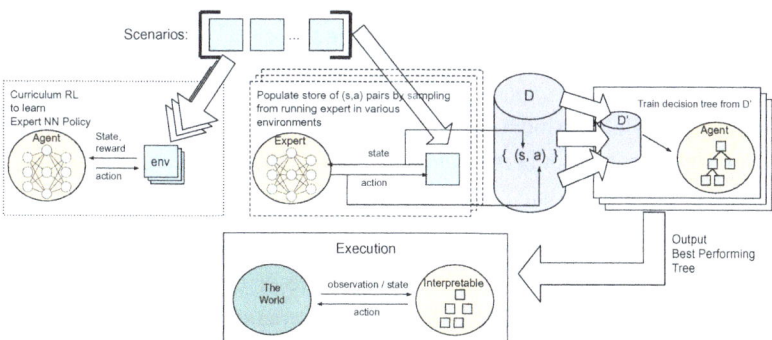

Fig. 3.1 In the MSVIPER Piperline [250], there are four parts of the process, the first three of which comprise the upper row of the image above. First, a black-box expert policy can be trained using different scenarios (environment configurations). Second, the black box is run in simulation and trajectories are generated which are stored in a database. Then in the third step a decision tree is trained using trajectories from different datasets. The best performing tree is output by the algorithm and can be used. In the fourth step (not shown in diagram), the tree can be modified to fix errors and improve the policy without retraining. A similar process occurs in [249].

attempts to note under what circumstances various subgoals are pursued, and represents this strategy estimate in a decision tree format (See Sect. 3.5.1).

3.1.2 Single Altered Decision Tree

> Self-Explainable; Global; Model-Specific; When-Produced: After; Explanation Type: Tree; Has been applied to robotics in simulation and real world; Knowledge Limits: Complete; Explanation Accuracy: Certain

This subcategory of "altered decision trees" deals with trees that have some sort of nonstandard decision tree format so they are not "pure" decision trees as one would find in a typical classification problem. These trees might have probabilities inside nodes or add linear equations to leaf nodes.

The most prominent such tree is a soft decision tree. A soft decision tree can also be learned by policy distillation [58]. In this soft decision tree, each branch node itself is a neural net, and the leaf nodes are actions. While this might sacrifice some interpretability at the lower level, the goal is to balance overall interpretability with the increased performance and accuracy achieved by mixing neural networks into the nodes. The higher-level structure remains an interpretable tree.

The NLDTs discussed in Sect. 3.1.1 can also be considered an alternate tree format.

3.1.3 Multiple or Combined Decision Trees

> Self-Explainable; Global; Model-Specific; When-Produced: After; Explanation Type: Mixture of Tree; No, has not been applied to robotics; Knowledge Limits: Complete; Explanation Accuracy: Certain

There are also RL methods that result in a mixture of trees such as [44, 309]. How interpretable a mixture or combination of trees truly is remains debatable. However, even if they cannot be immediately comprehended by humans, they can still be analyzed by machine and have relevance for safety. This is also a part of interpretability, since understanding and verification requiring a machine is still a step above full opacity.

3.2 Saliency Maps

A saliency map is primarily used to explain tasks for which the states include images or video in whole or in part. Typically, "salient" portions of the image (portions that have specifically influenced the action choice) are marked. (The techniques used to generate saliency information for images could be used to generate such information for other media as well, but the method of communicating them to the user would have to be different.) The modified or generated image can be used to explain local decisions for a specific state or can provide global information (for example, by indicating important pixels for a specific class or action in general [10]). Originally, saliency maps were used for explaining supervised learning classification methods using deep neural nets or convolution nets [271, 280]. As we explore below, they have recently been used in reinforcement learning including robotics applications.

A user study investigating the usefulness of saliency maps alone and compared to global summaries [133] is discussed in Sect. 3.5. In Sect. 3.5.4, [118] applies a feature-attribution saliency method to Unmanned Aerial Vehicles (UAV).

3.2.1 Post-Hoc Saliency Maps via Backpropagation

Post-Hoc; Both Local and Global; Model-Specific; When-Produced: After; Explanation Type: Pixels (individual features); Has been applied to robotics in simulation only; Safety-relevant

The first type of saliency map we explore is one that produces both local and global explanations. Wang et al. [323] uses deep reinforcement learning to cause an agent to learn how to play a racing game. The input is a pixel image. Backpropagation enables them to calculate the derivative of the output action with respect to each individual input pixel. This paper uses two Q-networks, one to estimate state values and the other to estimate action advantage. They use the technique from [280] for each individual network to get saliency information.

Luo et al. [186] applies a saliency map to a simulation of an aerial drone avoiding obstacles. In Fig. 3.2, we see an example of a local visualization from this paper.

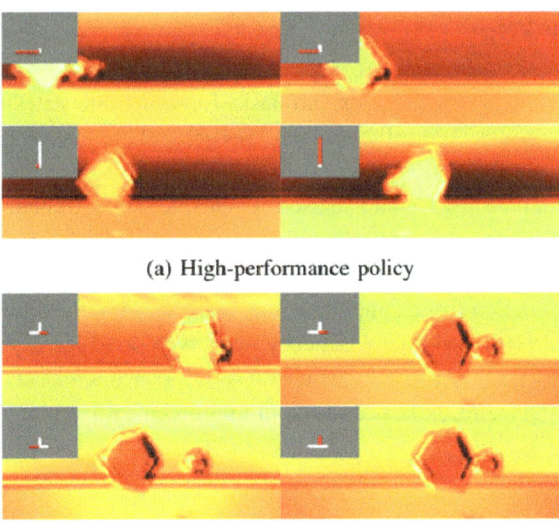

(a) High-performance policy

(b) Poor-performance policy

Fig. 3.2 In this visualization from [186], two visualizations of two different policies for a drone are shown, a (**a**) high-performance policy and a (**b**) low-performance policy. "The T-shape in the corner image visualizes action probabilities." In the original image, a red color indicates action taken (if the above image appears grayscale, this corresponds to a darker hue). Brighter hue in the image indicate a higher importance of those pixels. We see that in the higher-performing policy, the obstacles are marked as more important. This makes sense, since it is important to avoid them

3.2.2 Intrinsic Saliency Maps

Self-Explainable; Local; Model-Specific; When-Produced: During-Byproduct; Explanation Type: Pixels (individual features); Has been applied to robotics in simulation and real life; Legibility: Attempted; Readability: Attempted

While the above methods calculate saliency based on the neural net, methods in this subcategory generate saliency via more intrinsic means. There are several saliency map techniques that derive saliency information from the values of weights or other information in the neurons or networks. Some recent papers [204, 205, 219, 284, 340] applied techniques in this vein to the Atari 2600 set of reinforcement learning video game environments. They could also be applied to robots with cameras that learn using Deep Q-Networks (DQN) [199] or Actor Critic (A2C) [290]. (We discuss this further in the paragraph of Saliency maps in the discussion of future work in Chap. 5).

Although there has not been much work on using intrinsic saliency information in robotic learning applications, one interesting case is the use of a Recurrent Attention Model [284]

3.2 Saliency Maps

(which gives saliency/attention information for a Deep Q-Network) to facilitate Human Robot Social Interaction [239]. Qureshi et al. [239] uses a humanoid-like Pepper robot in an experiment where it is placed in a public area for 14 days and learns to interact with humans. The robot attempts to engage with humans and learns to recognize when a human might be interested by looking at the direction of the human's trajectory, the human's action, and other factors. It can attempt to engage with a human by looking at a human, hand waving, or hand shaking (reserving the latter for a nearby engaged human). Since the robot learns by focusing on the humans, the attention model highlights the humans in the images. This in turn is used by the robot to orient itself towards the human or to look at it. They compared a robot using the attention augmented model with a robot without it. They found that when the robot was more responsive to human stimuli (enabled by the attention module, which we consider a form of saliency), it resulted in a higher number of handshakes (which they consider to be indicative of higher human engagement with the robot). This is a good example of a robot using saliency information to improve its behavior.

A recent gradient-based saliency map approach integrates input perturbation saliency map techniques (see Sect. 3.2.3). Xing et al. [331] trains a perturbation saliency map using that map as a supervisory signal to train a gradient-based saliency map. They also use policy distillation [256] to guard against performance degradation and test on the CARLA autonomous driving simulator [79].

3.2.3 Post-Hoc Saliency Maps via Input Perturbation

Post-Hoc; Local; Model-Agnostic; When-Produced: After; Explanation Type: Pixels (individual features); Has been applied to robotics in simulation and real life

Input perturbation is a subcategory of saliency methods that does not require analysis of the model or taking a derivative, but that just runs experiments on the model with inputs of small variations. Controlling these inputs and measuring the outputs of the model demonstrate empirically which features are impacting changes in the model's output. This is beneficial because it allows for investigation and interpretation of a black box model.

It also addresses one of the issues with some of the previous saliency methods, which is that they might explain prediction by referencing pixels which are actually meaningless, meaning they are not always useful from a practical point of view [104].

Perturbation methods can choose to vary only those pixels or features that are specifically meaningful, and measure their impact on the output of the model, creating a saliency map of the result [89, 104, 336]. Empirically, the method discovers which features have the most significant effect on the output. Additional work to ensure that the saliency features/pixels identified are semantically meaningful is done in [111], which balances the effect that input change has on expected reward with specific relevance to the chosen action. This means that features that alter rewards of unchosen actions are considered less relevant than features that solely or primarily alter the reward of the action the policy has chosen for that state. Recent work has also conducted a benchmark study of four approaches to input perturbation saliency maps, albeit tested on Atari Agents [132].

Fujita et al. [92] takes the method from [104] and applies it to a manipulation robot in simulation and the real world. Nie et al. [218] takes the method from [89] and applies it to a robot swarm performing a navigation task in simulation.

Dai et al. [66] builds on a saliency map along with dimension reduction (Sect. 3.4.1) to provide explanations. In a task where a simulated robot arm must reach a target represented by a red circle, for example, the explanations can tell if the agent is using both the color and shape to recognize the goal or only the color.

3.3 Counterfactuals/Counterexamples

Humans communicate not just by explicit explanation, but also using examples or counterexamples. By sharing instances that are part of two separate categories (i.e., what the robot will or will not do under various circumstances) they can elucidate the nature of those two categories (i.e., describe the nature of robust behavior). There may be cases where coming up with examples and counterexamples is easier for a robot to produce than it would be to come up with a full explanation. Additionally, some information can be best communicated in this manner.

These types of "contrastive explanations" have been shown to increase trust and understanding of an autonomous agent. Chen et al. [54] performed a 100-person user study. They created a method for generating contrastive explanations of a robot agent's plan. They determined that providing these kinds of explanations improved the user's confidence in the robot, aided in their comprehension of the robot plan, and reduced the participants' "cognitive burden." Contrastive explanations can be supplied for planning over MDPs [288], but below we discuss counterfactuals, counterexamples, and contrastive explanations for reinforcement learning in robotics.

3.3.1 Counterfactual by Input Perturbation or Extra Information

Post-hoc; Local; Model-Specific; When-Produced: After and When-Produced:During-Byproduct; Explanation Type: state examples (images or text description); No, has not been applied to robotics.

Counterfactuals can be generated in multiple ways. One of the simplest, if one has access to a simulator, is to test different inputs and discover what changes they make to the outputs.

3.3 Counterfactuals/Counterexamples

This leads directly to an example that can be shown to a user [104]. Another method falling into the subcategory of input perturbation is when the examples and counterexamples are discovered by a General Adversarial Network (GAN) [222, 223].

In a different approach, [176] generates counterexamples during training based on extra information included during the training process. Specifically, they manually designed features that have semantic meaning. They model the problem as an MDP and, instead of a normal value function, used Generalized Value Functions [291], which describe how features will change over time in response to actions taken (as opposed to only how much reward will be received). If features are meaningful, then explanations can be generated from this process. One of their experiments involves a real-time strategy game where the agent generates explanations for why they chose to use resources to create certain units.

Surprisingly, these techniques have not been applied to robotics. While it might not be easy to apply in the real world, it can be done in simulation. (We should note that contrastive examples have been created for robots by a method in Sect. 3.3.2, and normal non-contrastive examples are produced by many methods such as the observation methods in Sect. 3.5).

3.3.2 Counterfactual by Model Checking

> Post-hoc; Local; Model-Specific; When-Produced: After and When-Produced: During-Byproduct; Explanation Type: state examples or text description; Has been applied to robotics in simple simulations.

Another series of techniques for creating counterfactual or contrastive explanations involves model checking, where a policy is learned involving a model that can be analyzed. Contrastive examples are derived thereby.

One such means is to use a causal model such as a Structured Causal Model (SCM) [230, 231]. An SCM is a directed acyclic graph that describes how variables affect outcomes. SCMs have many uses. If an SCM is provided ahead of time, a technique such as that in [234] can use reinforcement learning to learn a robot control problem over the causal model. This work creates counterfactual examples not for the purpose of explanation but to create additional training samples that are causally valid, improving performance. This could be extended to give explanations for humans as well. Madumal et al. [189] introduces an "action influence" model based on an SCM that seeks to approximate the causal dynamics of the agents environment with regards to the action the agent can take. This allows for the determination of why an agent took an action, and an explanation can be provided to a human to answer "why?" and "why not?" questions about agent actions. These researchers tested it on some common RL benchmarks and StarCraft. They also consider providing the user with a "minimally complete explanation," that is, the explanation that contains the smallest amount of information necessary to explain the decision. Madumal et al. [189] includes a user study with regards to the StarCraft environment that demonstrated how powerful these explanations were in helping humans understand and predict what the agent will do in the future, helping the human to understand the local strategy of the agent. One limitation of these methods is that the causal model must be provided, at least in part, ahead of time.

There are also ways to create counterfactual explanations without the explicit use of causality. In a robot gridworld navigation context, [54] calculates contrastive examples at critical states. The explanations are generated by analyzing the MDP that represents the world. For example, it will tell the user, "We move east at critical grid 10 because it leads to the shortest and most flexible future route." Not every action is explained, just the important ones (in this case, the point where the robot turns). They performed a user study demonstrating how these types of statements increased user understanding and trust. Policy Explainer [197] provides explanations for a policy by analyzing a tree structure. They create a policy using RL and then an explanation-generation procedure that is compatible with a tree style policy. (The paper itself did not go into detail on their tree generation process, how they converted a non-tree into a tree, or whether they learn the tree directly, but this survey has several methods for creating a decision tree policy with RL, as found in Sect. 3.1.1.) Policy Explainer answers "why" questions by traversing the tree structure and determining the decision rules that led to a certain outcome. It can answer, "Why action 1 and not action 2?" by comparing the two, calculating the difference in the decision rules that led to each, and sharing that difference with the user. It can also answer what the paper calls a "when" question, as in "when will you take action 1?" which it answers by calculating the circumstances (decision rules) in which that action would be taken. There is also a visualization aspect of the policy explainer discussed further in Sect. 3.14.

In the context of a user wanting a change of policy, [286] uses explanations involving counterfactuals to allow an agent to justify why the original plan is superior. In this framework, the agent has an expert policy, and the user might suggest an alternative policy in some respect. The system will explain why it prefers the original plan (if it does so) using contrast examples. Unlike other methods in this section, this does not require noting particular states or state sequences that could be critical or meaningful, and can be used when the state representation is not meaningful. However, the method does assume access to binary classifiers for custom user concepts.

In [95], safety criteria are incorporated with Bayesian Optimization to discover unsafe aspects of a trained policy (counterexamples) that must be corrected. In this manner, policies are made safer without significant loss in reward.

It is important for explanations to be simple for a human to understand, and [206] uses contrastive examples as part of a framework to aid in human understanding of robot explanations. In a simulated UAV package delivery scenario, a user can pick UAV's to assign to deliver packages, and the UAV can give explanations about behavior related to unexpected situations such as expecting a package for pickup and not finding it or finding a different package, or by responding to an assignment by saying it cannot pick up a package because it is charging.

3.4 State Transformation

One category of tools that can be used as standalone methods or in combination with other methods is that of state transformation. Robotics applications often have a very large number of input elements from sensors such as cameras and lidars or other raw state representations. Also, it can be costly for a robot to take actions, so anything that speeds up learning is helpful. We discuss two subcategories. The first is in Sect. 3.4.1 Dimension Reduction, where many dimensions are converted into fewer dimensions, enabling easier understanding. Section 3.4.2 describes various types of meaningful representations into which raw states could be converted with improved explainability.

State Transformation methods are also interesting in combination with other techniques. Even a state transformation method that is not directly meaningful can enable other explainable methods that are (for example, using a Variational Autoencoder (VAE), a state transformation method, can enable explainability in Hierarchical Reinforcement Learning (HRL) via primitive generation [332, 333]).

This paper discusses only state transformation in reinforcement learning. A survey of more classical and recent approaches up to 2018 can be found in [167].

3.4.1 Dimension Reduction

> Post-hoc or Self-Explainable; Global; Model-Agnostic; When-Produced: After or When-Produced:During-Intrinsic; Explanation Type: simplified/abstract states, sometimes in the form of a simplified/abstract MDP; Has been applied to robotics, in simulations only.

A simple way to reduce the size of the state space is to cluster states in the raw state space into an abstract state and an abstracted space. In [8], the raw states are clustered into K abstract states. When new raw states are visited, the cluster centers might change. When a raw state is closest to one of the K abstract states, the agent will take the action associated with that abstract state. In [9], they take a slightly different approach for state abstraction, generating a policy that is like a program, with a sequence of IF-THEN blocks, where the THEN component is an action, and the IF is a condition upon the state space. (Therefore, it aggregates like a decision tree does, although with the explicit concept of state aggregation as a goal instead of a side effect.) Again, there are K clusters, and the agent takes an action associated with the cluster closest to the input state. This method also includes a default action if the state is too far from all clusters. They test this in simulation on half-cheetah [42] and Ant environments [262, 300]. (The programmatic nature of this method is reminiscent of Symbolic Learning, Sect. 3.16.3).

Other methods explicitly create an abstract MDP as a means of simplifying a more complex, known or unknown MDP. An approach that creates such a Semi Markov Decision Process (SMDP) is [356]. They utilize t-SNE [307], which is a way to visualize multiple dimensions by transforming and reducing them to two or three dimensions. It is especially useful for data that has high dimensions lying on different but related manifolds ([307] gives the example of camera images of the same objects from different viewpoints). In [356], the authors first train an agent on an environment using DQN. Then they evaluate it by running the agent and recording visited states, neural activation, and Q-values. This provides enough information that t-SNE can be applied. There is a further clustering step to cluster the data in this lower dimensional space. Finally, they attempt to create an SMDP over these clustered states, estimating the transition probabilities and reward values. Now we have an abstract MDP that can also be easily visualized. An example for the Atari game Breakout is shown in Fig. 3.3. A similar work [341] performs a similar process by taking a hierarchical approach, where the higher-level policy is abstracted in such a manner. One potential danger to be aware of when utilizing the reward to help aggregate states is that perceptually different states might be grouped together because of similar rewards [199]. Topin et al. [302] is a further approach to generating an abstract MDP. Here, the MDP represents the policy. Given a large continuous MDP, [25] reduces it into a smaller MDP that can be planned over in a hierarchical way.

3.4.2 Meaningful Representation Learning

Self-Explainable; Local, Global, or Either; Model-Agnostic, Model-Specific, or Variable; When-Produced: During-Byproduct or When-Produced:During-Intrinsic; Explanation Type: Representation (sometimes, symbolic plan upon the representation); Has been applied to robotics, in simulation and real life; in the case of graph representations of visual input using visual entity graphs, Explanation Accuracy is Certain

This subcategory describes methods that create or use a representation of the state space that increases interpretability. Representation learning attempts to transform a raw state space into a learned abstract state space where the features of the learned space better characterize the important aspects of the space and can be more effectively and directly used [31, 167]. The learned features, when they have low dimensionality, can improve generalization and decrease training time. When these learned features are in whole or in part semantically meaningful, we propose the term Meaningful Representation Learning, since the learned space aids in human interpretation as well.

The first method we include, found in [177], is similar to the MDP methods in Sect. 3.4.1 but does not create an MDP. It is similar to clustering states but specifically creates a graph format (although not an MDP graph). The graph created is similar to the graph made by the

3.4 State Transformation

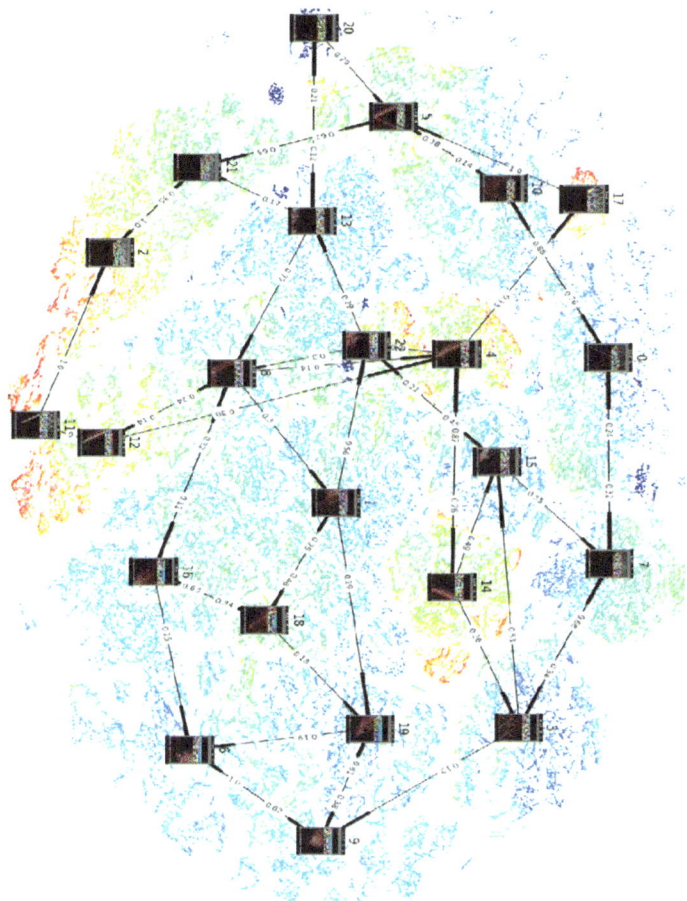

Fig. 3.3 In this image from [356] (rotated 90° to fit on the page), we see a Semi Markov Decision Process created and diagrammed for the Atari game breakout. Similar states are grouped together into clusters, serving as nodes; the nodes are described here by images. In each image (the mean of the cluster), the ball location and direction are shown providing a description of the circumstances for a given cluster

Symbolic Planning Over Graphs methods (Sect. 3.4.2), but it does not use symbolic methods. Lippi et al. [177] creates a graph in the latent space, clustering states and embedding images. An agent can then plan over this graph, generating sequences of images to form a plan and creating an action plan for lower-level actions to get from one node to another. The graph and the plan across it represent interpretability. They test this on a simulated robot block stacking task and a real-world t-shirt folding task. A diagram of the t-shirt task plan is found in Fig. 3.4.

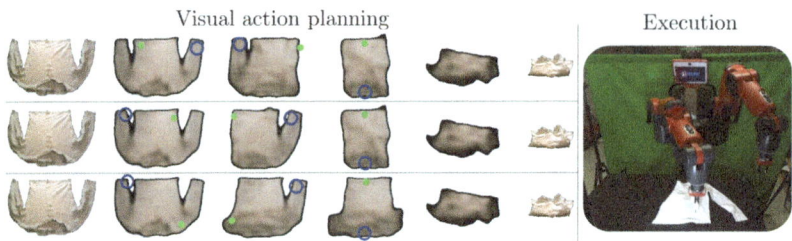

Fig. 3.4 In this image from [177], a plan is shown for folding a shirt. Clustering in the latent space creates distinct "steps" for the plan, and embedding images allows for showing these images. Now just the lower level must be solved to get from one step to the next instead of needing to solve the task in its entirety

Graphs can be used to provide insight in a totally different way as well. In another technique that involves symbolic learning (Sect. 3.16.3) and meaningful representation learning, [274] uses Visual Entity Graphs to represent the image input. If a robot is performing a task involving visual input, that visual input itself can be represented as a graph. Sometimes, an agent will care more about the relationship between physical objects than the specifics of where the objects are, e.g., understanding what object is on top of another one. The graphs used in this paper are hierarchical. Nodes represent visual entities such as physical objects and edges represent their relative spatial configuration. The entities represented by the nodes will persist and the end relationships may change over time. This type of representation also provides robustness to a camera looking at things from different points of view. A human's hands also count as an object-node in the graph. In this paper, they use this representation to perform a learning-from-demonstration task. With this technique, they can learn simple tasks from a single demonstration with a real Baxter robot: pushing an object, stacking an object (putting one object on top of another) and pouring (pouring liquid from one container to another). Figure 3.5 is a representation of this. An example of the further usefulness of this representation is found in [175], which utilizes a graph representation as input for a graph neural net learning method that uses symbolic reasoning to plan. See more discussion of it in Sect. 3.16.3.

Another type of symbolic meaningful representation learning is found in [343], which learns the logical relationships between entities. They use an attention mechanism to combine relational learning, reinforcement learning, and inductive logic programming [207]. States, actions, and policies are represented using predicates, which allows a relational language to describe them. A byproduct of the use of this relational language is that a human can understand the robot's perception better. One limitation of this approach is that the entities must be chosen ahead of time.

The next class of methods to create meaningful representations uses human sourced metadata, feedback, or domain knowledge. Sodhani et al. [282] uses metadata and shared skills across multiple tasks to earn interpretable representation. They focus on the idea

3.4 State Transformation

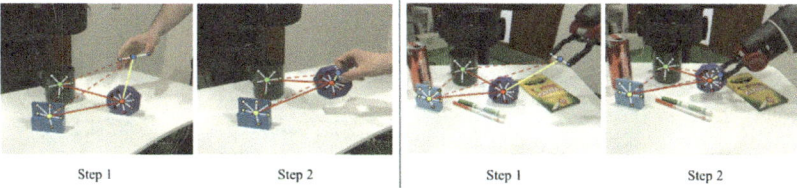

Fig. 3.5 Graph structured visual imitation from [274]. The human demonstrates a simple motion. A graph describes the relationships between the objects, serving as a meaningful representation. The robot learns how to perform the same task from the single demonstration using this representation, which also serves to provide interpretability to a human

that different tasks might have commonalities despite different state spaces. Similar objects might appear in different state spaces and similar actions might be available in different action spaces. They create encoders that focus on specific objects or actions, and rely on metadata to describe the situation at hand and choose what is being learned or utilized. They test on a robotic relations simulation in Meta World [338]. Another way human knowledge can be incorporated is when humans encode their knowledge and loss functions or unsupervised and self-supervised network, as in [40, 41]. These papers apply this concept on a multi-target navigation domain in simulation. The insight in this approach is to say that changes in reward between law states should correlate with reward changes between corresponding simplified states, a property they call "reward proportionality." They create a simplified state space by attempting to satisfy this property. The method clusters together states with similar reward variations (focusing on reward magnitude) independently of the kind of action taken. They also use a similar prior/loss function they call "reward repeatability," which seeks to group states by reward direction (distinct from magnitude).

The next set of meaningful representation learning methods that we discuss involves algorithms that autonomously discover and define their learned latent states. Raffin et al. [241] decouples feature extraction from policy learning. They train with a reward function that has a few components. One component is for learning a task-independent autoencoder over the states. Another component is for learning a task-specific reward. These are combined to develop the latent space, and they test it on a sparse reward simulated robot arm task. Instead of a normal autoencoder, [110] uses a similar technique but with a Variational Autoencoder (VAE), a component rapidly gaining popularity [147, 244]. They apply this to simulations on an autonomous car simulator. In [203], a VAE is combined with a contrastive estimator. They test it on an OpenAI gym Atari Racing Environment. With a little care, [90] demonstrates how to use a latent space to gain verifiable behavior. They use an environment (Atari Mountaincar) where the raw space is raw pixels. They learn a latent representation where the features correspond to the movement of the car and linear dynamics apply (as they do not in the raw space). Since linear dynamics apply here, the authors can achieve theoretical guarantees on the behavior of the car agent. A VAE is used to create a latent space for a real-world robot manipulation task in [243, 348]. Nair et al. [212] extends this

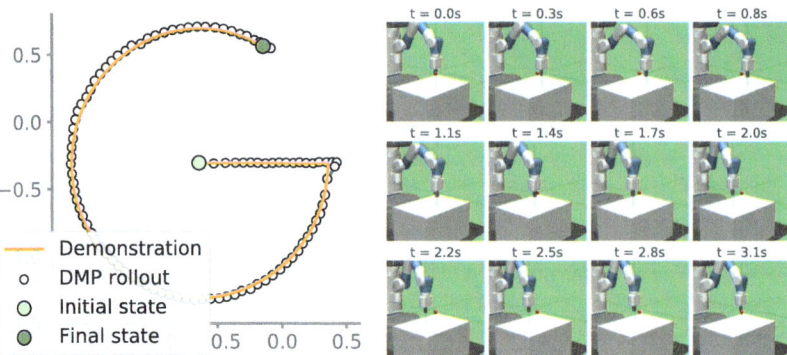

Fig. 3.6 The Fetch3D robot arm manipulation task from [140]. At right, see a sequence of 12 images in the form of the raw state space as the arm moves. At left, a plot of the first two dimensions of the learned meaningful latent space

approach to skill learning, imagining goals and proposing goals that the learner can use to learn skills in the latent space. Using these goals in the latent space helps ensure that it is more likely that they are relevant and useful goals than if a random sample were to be taken from the raw state space. A similar latent state representation is used by [350] for meta learning (multitask learning), training a robot in the real world to insert an ethernet cable into a slot. NewtonianVAE [140] has a goal of learning representation that converts from raw images to a state representation that is compatible with traditional Proportional-Integral-Derivative (PID) control [328]. They tested this with three simulated continuous control environments, including arm reaching and Fetch3D, where a robot arm manipulates objects. This method is interpretable to the same degree as traditional planning due to the conversion. Consider also Fig. 3.6.

State transformation is also used in the constrained learning method [82], described in Sect. 3.8.

3.5 Observation Based Methods

An observation-based method is one in which the interpretability is derived completely from observing a policy in operation. It is very useful because it is extremely model-agnostic.

Also, many previously discussed methods focus on a specific local decision, or describe the global policy with an interpretable structure. As [12] points out, it is desirable to have a means for producing a *summary* of the robot's strategy that can be presented to a human. Observation based methods are one way to achieve this.

3.5.1 Observation Analysis: Frequency or Statistical Techniques for Policy Understanding

> Post-hoc; Global; Model-Specific or Model-agnostic; When-Produced: During-Byproduct; Explanation Type: videos and/or trajectories of state-action pairs (or state-action-Q-value, or state-action-additional-info tuples); No, has not yet been applied to robotics.

This subcategory involves techniques that calculate statistics related to observations to develop an understanding of an otherwise black box policy.

During the training of a policy in [265, 266], video is recorded of "important" trajectories. Further video is taken after the policy is fully trained. The goal is to take video of trajectories that seem important or illustrative of agent decision making, which can be shown to a human to convey a general understanding of what the policy does. Important actions are those with the largest difference in Q-value between the chosen and not-chosen actions. Important trajectories are those with important actions. This work also tracks the frequency with which various types of situations arise and can communicate that to the user. Similarly, [11, 130] show humans the most critical states, where taking different actions would have the most significant eventual impact. This is very model-agnostic, but the explanation is imprecise.

A similar type of summary is performed by [14] to explain not a policy in isolation but the differences between different agents (since such differences might not show up in a summary by default).

While [11, 14, 130, 265, 266] use different technical means to achieve a similar conceptual idea, [2] goes much further, seeking to estimate a model of the environment (transition dynamics) from recorded trajectories. They calculate additional insights from the data as well. If a trajectory leads towards a human-designated subgoal, the algorithm may assume that is the goal of that trajectory, adding this information as a "strategy label." This allows for looking at what parts of a policy are used to achieve each subgoal. They can also determine what subgoal is likely to be pursued based on the labeled features and represent this by a decision tree. This technique has limited applicability since it requires some human engineering.

The type of explanations described here can be called "global summaries." A user study regarding the effect of global summaries vs. saliency maps is conducted in [133]. As they note, global summaries demonstrate important trajectories of states to give understanding about the overall policy, and saliency maps (Sect. 3.2) describe what specific information an agent is attending to regarding a specific local decision. Global summaries with important or critical states shown to the user increased user understanding compared to summaries with randomly chosen states. The results were more inconclusive regarding the usefulness of saliency maps, whether alone or in combination with global summaries.

3.5.2 Observation Analysis: Human Communicative Trajectories for Goal Understanding

> Post-hoc; Global; Model-Specific or Model-agnostic; When-Produced: During-Byproduct; Explanation Type: trajectories; Has been applied to robotics in simulation only; Predictability: Attempted.

Like methods in the previous subcategory, [160, 161] collect and showcase trajectories to humans. The purpose here is not to describe the overall strategy but rather to describe what the agents' goals might be or for what reward function the agent might be attempting to optimize (if the human does not know). In general, this type of work can also be called a "summary," although it can also be referred to as "model reconstruction" [10]. The goal of [131] is to highlight the different situations that correspond to different weights in the various parameters in the reward function. One of the domains this paper uses is an autonomous car environment. Their reward function could choose to more heavily promote safety or efficiency. In the same scenario, a more safety-oriented car might choose to stay in the same lane whereas a more efficiency-oriented car might choose to attempt to pass other cars. Thus, from observing the type of action the car takes in such a situation, a human can have an idea about what kind of parameters were more important to that agent. They also have a user study that bears out this reasoning regarding human understanding. See a visual description in Fig. 3.7. This kind of method of course requires a specific kind of reward function and so is a bit model-specific, but if such a reward function is constructed, it is otherwise agnostic. In [129], critical states are again shown to the user to increase trust and the predictability of the model.

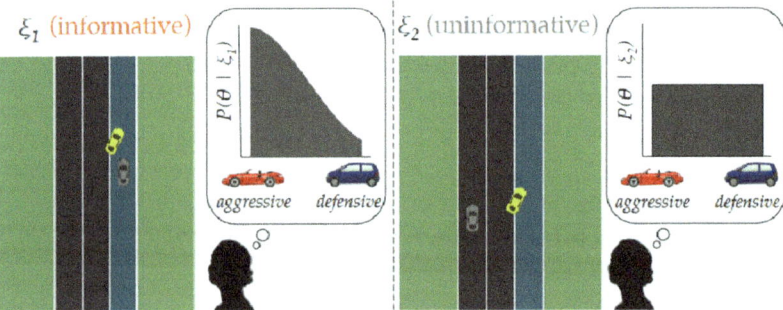

Fig. 3.7 Two examples of states from trajectories a user could be shown as part of a summary from [131]. The example on the left is more communicative than the example on the right because it demonstrates to the user the aggressive nature of the efficiency oriented policy

3.5.3 Observation Analysis: A/B Testing

Post-hoc; Global; Model-Variable; When-Produced:After; Explanation Type: hidden policy variables; Has been applied to robotics; Knowledge Limits: Partially Attempted

The method in [267] tries to find causal relationships by observing the behavior of the agent to explain why state transitions are not as expected. They test on a ride sharing planning algorithm domain where drivers are given recommendations and can then choose to follow them or not. The agent attempts to learn how to make recommendations that will achieve its efficiency goals, which means also making recommendations that will be followed by the humans. The algorithm considers possible policy improvements and uses A/B testing [5, 281] to determine which of the possible improvements is better. The idea is that the recommendation the driver chooses is affected by the recommendation itself and by hidden factors that this method (POMEE-U) attempts to infer. The authors intend their work to be a general pipeline for optimizing reinforcement learning policies in the real world using historical data. They observe the transition dynamics and build the model and simulation. In this case, they model the drivers, the ridesharing platform, and the hidden variables. They attempt to predict what the drivers will do in response to the agents' recommendations and then explain the differences that occur. The explanations here are the information gleaned from the hidden variables in the environment.

We note in the subheading that this partially attempts to define Knowledge Limits (Section 2.3) for the policy. While the paper cited does not formally allude to any such discussion, we note in our evaluation that this method could be seen to investigate the limits of an algorithm and correct it.

3.5.4 Training Data Observation Analysis

Post-hoc; Local; Model-agnostic; When-Produced: After; Explanation Type: feature weights; Has been applied to robotics in simulation only

The interactions between the agent and environment are again observed in [69] to develop a state transition model. A human defines an objective or risk function that takes as an input the state or states and outputs the riskiness of that state. This determines risk and can be used in explanations. For example, they use a "Bipedal Walker" environment. They defined a "risky" state as one that would cause episode failure or one for which there was no known policy that would not result in a failure state. The Bipedal Walker is a legged robot, and this method allows for determining that the two most important features are the "angle of joint between first leg and the body and its speed" [69]. This is because, in a "risky" situation, the value for the risk metric was higher with respect to these two features than for the rest of the features such that "increasing the angle and speed of the front leg, given the current state the agent is in, will result in a risky situation."

SHapley Additive exPlanations (SHAP) have been widely used in supervised learning to explain outputs via feature attribution [185]. SHAP uses insights from game theory to do this. After training a policy, the inputs and outputs can be observed, and the network inspected according to the SHAP algorithm. For each predicted output, each feature receives an importance value. Knowing which features are important is a form of explanation if the features are semantic. SHAP methods are beginning to see use in RL as well. In [174], this technique is applied to a driving simulator, where are car must complete a route as energy-efficiently as possible while maintaining speed and acceleration limits. Liessner et al. [174] shows how, at a given point in time, we can see the degree to which speed limits or other factors affect the decision the agent made. Jiang et al. [141] and Rjoub et al. [246] are more recent autonomous driving applications that use SHAP for explanations. He et al. [118] applies this technique to a learned UAV path planner, and extends SHAP to apply to visual features (so it's a type of saliency method as well). The explanations are tested in simulation, but the planner itself is tested on a real-world UAV. Løver et al. [184] applies SHAP explanations to maritime autonomous docking.

3.5.5 Interrogative Observation Analysis

> Post-Hoc; Local; Model-specific; When-Produced: After; Explanation Type: Highlighted Trajectories; Has been applied to robotics in simulation only

Instead of or in addition to getting a general sense of important trajectories for a policy, sometimes a human will want to specifically search for a type of trajectory. RoCUS is framework for explaining learning methods where a user specifies a type of trajectory or situation they want to look for and the method searches for a type of trajectory with respect to a particular learning method across multiple environments [354]. For example, a user can learn why a navigation robot is not moving in a straight line or why a manipulation arm robot may be moving an end effector near the goal but not to the goal.

3.6 Custom Domain Language

> Self-Explainable; Global; Model-specific; When-Produced: Before; Explanation Type: Language; Has been applied to robotics in simulation only

Programmatically Interpretable Reinforcement Learning (PIRL) [312, 313] is a form of RL where the policy itself is represented by a level domain-specific programming language. Like the policy distillation decision tree methods from Sect. 3.1.1 [249, 250, 252], PIRL first learns a neural net policy with Deep Reinforcement Learning (DRL) and then builds a new policy in the language space, using the expert policy to guide a search over the space of possible policies. Just like the neural net policy, the policy created by the programming language attempts to maximize reward. Restricting the components of the policy to this domain-specific dataset results in an interpretable algorithm. The authors successfully use this technique to succeed in a simulated autonomous driving task. A learned programmatic sub-policy for the subtask of acceleration is given in Eq. 3.1.

$$\begin{aligned}
&\textbf{if } (0.001 - \textbf{hd}(h_{TrackPos}) > 0) \textbf{ and } (0.001 + \textbf{hd}(h_{TrackPos}) > 0) \\
&\quad \textbf{then } 1.96 + 4.92 * (0.44 - \textbf{hd}(h_{RPM})) + \\
&\quad 0.89 * \textbf{fold}(+, h_{RPM}) + 49.79 * \textbf{hd}(\textbf{tl}(h_{RPM})) - \textbf{hd}(h_{RPM}) \\
&\quad \textbf{else } 1.78 + 4.92 * (0.40 - \textbf{hd}(h_{RPM})) + \\
&\quad 0.89 * \textbf{fold}(+, h_{RPM}) + 49.79 * \textbf{hd}(\textbf{tl}(h_{RPM})) - \textbf{hd}(h_{RPM})
\end{aligned} \quad (3.1)$$

where h_{RPM} and $h_{TrackPos}$ are histories of sensor readings, **fold** is the standard higher order combinator $(f, [e_1, ..., e_k], e) = f(e_k, f(e_{k-1}, ... f(e_1, e)))$, **hd** "returns the element in an input sequence representing the most recent timepoint," and the policy itself and the constants in it are learned autonomously by the method after these and other operators have been defined.

PIRL is limited by the requirement of finding or creating the domain-specific language, but once created, the expressive insight it yields for a human is valuable.

3.7 Constrained Learning

> Self-Explainable; Global; Model-specific; When-Produced: Before; Explanation Type: Rules or other constraints; Yes, has been applied to robotics in simulation and real world; Explanation Accuracy: Uncertain but Measured; Safety-relevant

Rule-interposing learning (RIL) [345] embeds high-level rules into the deep reinforcement learning for a task. They use a Deep Q-Network (DQN) method combined with a set of rules. At each timestep of the learning process, after evaluating the Q-values for possible actions, the method checks if any rules are applicable. If so, it might follow the rule according to a certain probability that depends on the stage of learning and the properties of the rules themselves. This paper tests the method on gridworlds, breakout, and other non-robotics tasks. It discovers that these heuristic rules accelerate learning. More importantly, the safety-relevant rules guard the DQN learning agent against unsafe learning.

Constrained Policy Optimization (CPO) is a type of reinforcement learning where the learning is mathematically bounded in a way intended to enforce safe actions during learning and execution [3]. Anderson et al. [16] improves upon Constrained Policy Optimization by projecting the state into a symbolic space that can be analyzed and performing safe updates there. Yang et al. [335] builds on CPO for safe legged locomotion. Kobayashi [149] combines constrained learning with a local Lipschitz continuity constraint (to ensure smoothness) at each timestep for smooth safe robot movement in simulation. Thumm and Althoff [299] uses constrained learning in a simulated robot manipulation task to ensure that the arm always stops before it would impact a human with which the robot is coworking. In [59], safety information is specified by a human ahead of learning to constrain it in a mapless navigation task.

This category of methods has clear implications for robotics, where safety is a massive concern.

3.8 Constrained Execution

> **Uncertainty-Aware Constraints**: Post-hoc; Local; Model-specific; When-Produced: During-Byproduct; Explanation Type: model uncertainty; Has been applied to robotics in simulation only; Knowledge Limits: Partially Attempted
> **Planning Constraints**: Self-Explainable; Global; Model-specific; When-Produced: During-Byproduct; Explanation Type: plannable latent representation; Has been applied to robotics in simulation only

In contrast to constrained learning, constrained execution involves processes that restrict or modify a policy's output after the policy has been otherwise fully learned. The authors of [187] use uncertainty to inform the constraints. Lütjens et al. [187] trains an ensemble of methods on different parts of the observation space. One of the goals of the method is not just attempting to predict the best action but to measure the agents' own degree of certainty about whether it is the right action. Although they do not use the term "knowledge limits," this is an example of incorporating knowledge limits into learning. Uncertainty is estimated by comparing the predictions for the different methods in the ensemble. If predictions differ wildly, the agent is considered to be more uncertain, and if the predictions agree more, the

3.8 Constrained Execution

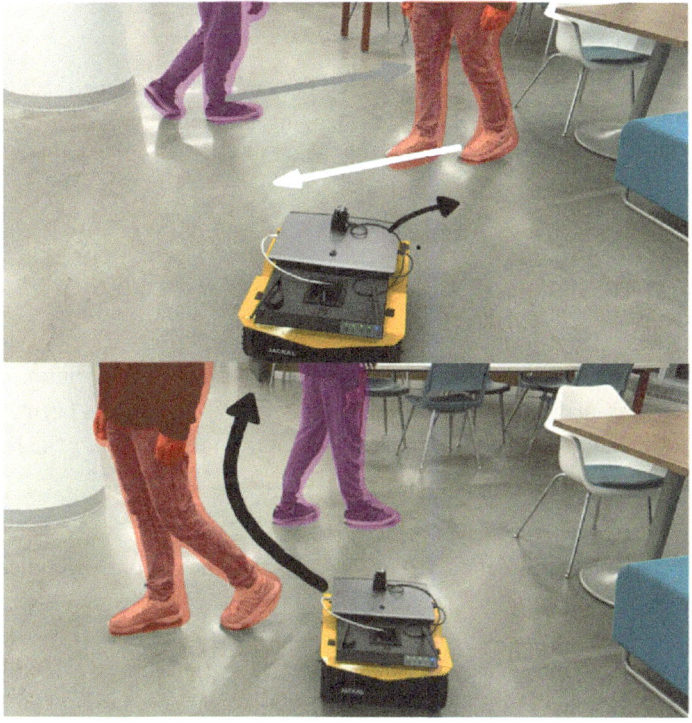

Fig. 3.8 In this image from [227], the lighter-colored arrows indicate predicted human paths, and the black arrow indicates a safe robot path.

agent is considered to be more certain. They apply this to a navigation task and predict a collision probability as a result of action choices. If the agent is more uncertain, it will give more leeway to obstacles. This is tested in simulation. A different uncertainty method with a similar purpose is demonstrated by [294] where they apply it to the CARLA autonomous car driving simulator.

It's also possible to use motion planning itself to constrain the actions. This is the approach chosen by [155], which calculates safe actions with motion planning and ensures that only actions that pass this filter are chosen and executed. They demonstrate this on a car in a lane-changing task. This type of approach is applied to a real-world navigation robot with DWA-RL in [227]. See a picture in Fig. 3.8.

A simulated robot box pushing task is tackled in [82]. A latent graph space is learned. A human can specify safety constraints by providing demonstrations that either do or do not satisfy the constraints. These constraints are enforced as part of planning over the abstract graph that is created. The abstract graph is a learned representation, making this method also within the category of State Transformation: Meaningful Representation Learning (Sect. 3.4.2).

This survey considers the constraints themselves to be a form of explanation.

3.9 Hierarchical

Hierarchical Reinforcement Learning (HRL) has been extensively studied [228]. Many have used it to improve performance, tackle challenging problems, or enable transfer learning [228]. In this category, we talk about the benefits of certain hierarchical approaches with regards to interpretability and explainability.

3.9.1 Hierarchical Skills or Goals

> Self-Explainable; Global; Model-specific; When-Produced: During-Intrinsic; Explanation Type: task plan, symbolic structure, or simplified MDP; Has been applied to robotics in simulation and real world; the "High Level Interpretability via HRL" collection of methods has Certain Explanation Accuracy

Hierarchical learning provides an elegant way to bypass the performance-interpret-ability trade-off. When there are multiple levels, different levels can provide different levels of interpretability. For many applications, higher-level interpretability is very beneficial even if lower-level policies remain black box. We call this "High-Level Interpretability via HRL."

This approach is taken in [273], where skills can be composed in a plan. The higher-level plan is human interpretable, and lower-level skills are black box. Lower-level skills are learned by attempting to achieve sub-goals. They tested on a block stacking simulation. One of the drawbacks, however, is that the sub-goals must be at least partially created by domain experts. In the "dot-to-dot" approach from [34], a robot arm chooses its own subgoals to move to for a manipulation task. Since it proposes its own subgoals, which it uses to train the lower-level skills, it does not require a human expert. However, this type of self-learning might not work for all domains. It requires a space for latent space where random sampling of such subgoals yields good results. They tested it on a MuJoCo-based model of the Fetch Robotics Manipulator as well as a Shadow Hand. While [34, 273] are demonstrated in simulation, [191] extends this approach to a real-world pick and place task. Training is initially done in simulation, and human knowledge is used to define sub goals. The robot learns skills from these sub-goals, which can be used at a higher level to solve the task. An additional benefit of the interpretability is demonstrated by [120], where the lower-level policies are environment- or robot-specific, and higher-level policies can be transferred between robots with different form factors. A similar insight is brought by the Transferable Augmented Instruction Graphs (TAIG) concept from [248].

Hierarchical primitive composition involves applying multiple levels of hierarchy with nested primitives, potentially yielding interpretability benefits [166]. Lee and Choi [166] also accounts for different skills having different action spaces and applies it to a simulated robot pick and place task.

3.9 Hierarchical

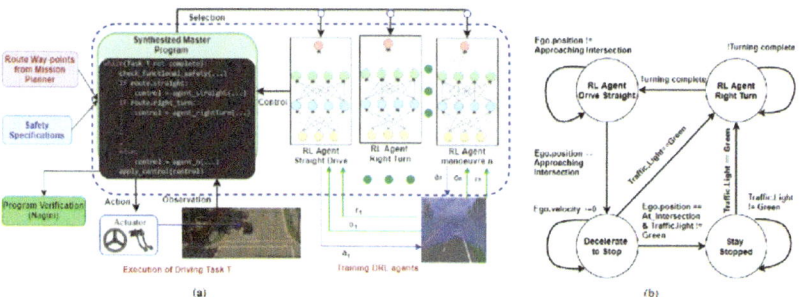

Fig. 3.9 A diagram of a hierarchical learned approach for car driving from [96]. The higher level is interpretable, and safety constraints can be added. The description of this diagram from [96] is: "(**a**) The right side of the figure depicts the asynchronous training of RL agents with individual observation/action space and reward signal. The left part depicts the program-controlled execution of a driving task where the program checks for functional safety and triggers RL agents as per route plan generated by the mission planner or as per control strategy recommended by safety module. (**b**) State machine for right turn manoeuvre at a 4-way traffic light controlled intersection without any dynamic object"

An additional benefit of the multilevel hierarchical approach becomes apparent when we introduce symbolic learning. In [334], the agent learns a high-level abstract graph as the high-level space and then uses a black box model for the reinforcement learning to train options to serve as transitions between the nodes of this graph. Different nodes in the abstract graph have symbolic differences. A symbolic planner can be used to choose how to move around the graph, while the low-level policies solve the sub tasks and find options to move between the nodes. Where the raw environment is too complex to gain a transition model (which is often the case in robotics environments), the simplified graph may be sufficiently finite to learn a transition model. In [334], they test it on an Atari game, Montezuma's Revenge, and a multi-room exploration game. For Montezuma, they use predefined nodes. A transition model can be learned at the high level, and options learned at the lower level. Gangopadhyay et al. [96] does this for an autonomous car racing simulation (CARLA). They learn multiple low-level skills for autonomous driving as the lower-level options. The higher-level policies use these and plan a high-level symbolic policy over them. See a diagram of this in Fig. 3.9. Another approach to using planning over a graph as a higher level policy is found in [143], where the graph seeks to subdivide the state space according to a specification (model of the problem constraints).

Several HRL methods involve a higher-level graph or MDP to which symbolic methods can be applied, and these are discussed in Sect. 3.4.2 since the graph results from a state transformation.

3.9.2 Primitive Generation

> Self-Explainable; Local; Model-specific; When-Produced: During-Intrinsic; Explanation Type: skill choice; Has been applied to robotics in simulation only; Explanation Accuracy: Certain

This subcategory deals specifically with HRL methods that involve generating the primitives or skills that comprise a lower layer and get used by a higher layer.

Similar to some of the previous methods in this category, [127] has lower-level learned black-box skills that a higher-level policy uses, but the focus is on analyzing the high-level policy to identify skills that are missing and then training those skills. They test on a simulated robot in a maze. Jiang et al. [142] learns skills that can be composed (not just used individually). Specifically, it uses language itself as the abstraction layer (so it could also be considered instruction following, as in Sect. 3.15). They test with a quasi-robotics physics task. A simulated manipulation task is tackled by [173], where an agent learns how to solve more complex tasks by composing solutions to smaller tasks. In [332, 333], a variational autoencoder is used to project the original state space into a meaningful low-dimensional feature space. They create representative state primitives that are then used. They sample from the feature space to create skills to learn different tasks. When learning a new task, the feature space is constrained to be close to the previous feature space, so that previously learned tasks can be protected. In other words, diverse tasks are cast into similar state space representations so that similar skills can be used across them.

An approach to skills learning involving a symbolic higher-level policy, Bilevel Neuro-Symbolic Skills, is discussed in Sect. 3.16.3.

It is also worth noting that not all primitive generation involves interpretability. For example, [170] learns the high-level and low-level skills simultaneously (as do many HRL methods). This is good for learning, but the lack of constraints on the nature of the skills thus learned and boundaries between them and between levels of the hierarchy mean that the learned skills may not have any semantic meaning. Lacking such, the high-level policy is still a black box and is as interpretable as a normal neural net, even if the performance is improved. Not all HRL or skill generation yields interpretability. We have featured some that do, whether as an explicit goal or without having it as a goal. There are several significant approaches that focus specifically on learning diverse skills, such as [84, 346], but these have the same lack of interpretability issues as [170] and others.

3.10 Machine-to-Human Templates

Humans communicate best using natural language, and thus researchers will sometimes choose to create agents that try to communicate with the human using natural language [70]. This can result in great interpretability when successful. In this category, we discuss template-based approaches for generating or parsing natural language explanations.

3.10.1 Model-to-Text or Policy-to-Text Templates

> Post-hoc; Local; Model-specific; When-Produced: After; Explanation Type: text; Has been applied to robotics in simulation only; Knowledge Limits: Complete; Explanation Accuracy: Certain; Safety-Relevant

Model-to-Text or Policy-to-Text Templates refer to methods that use templates to convert between an MDP-based RL policy and human language or between the MDP itself and human language. One example of this approach to generating text explanations that algorithms use to increase trust is from [322]. They describe it as "domain independent," although this is technically only true after the template is created (since creating the template may require domain knowledge, although one template could be created for different domains). They take a POMDP with known transitions and beliefs and use a template that defines how to convert that information into language. They test this on a simulation involving a human-robot team performing search and rescue in a counterterrorism scenario. The robot attempts to also give confidence intervals about its knowledge, making this one of the few methods that incorporates Knowledge Limits, although without using that term. The robot describes percentage confidence and indicates which sensors collected what data that led to its described beliefs. Therefore, although there is some effort upfront in terms of having created a template, benefits for the user are demonstrated. Maes et al. [190] utilizes an algebraic-formulated human readable language to perform conversion. First, it selects a human readable set of operators and terminal symbols, defines a metric to evaluate the language from these operators and symbols, and then applies the metric to evaluate the interpretability of the policy. Genetic algorithms are used by [119] to turn a policy into linear equations with if/else rules that can be expressed in natural language. The constraints of the policy make it difficult to learn in a normal manner, and thus the genetic learning is used. Because the policies are constrained to use these if/else operators corresponding to natural language, whatever policy results will be naturally understandable.

3.10.2 Query-Based NLP Templates

> Self-Explainable; Local; Model-specific; When-Produced: After; Explanation Type: text responses to query; Has been applied to robotics in simulation only

In some situations, we might not want the robot to volunteer everything it knows for the sake of time. Instead, we want the ability to interrogate the agent. Such a problem is tackled by [116], which provides a template-based method for answering queries for tasks that have an MDP solved by Q learning. This method answers the following types of queries: (i) What type of states and conditions would result in performing a specific action? (ii) List the actions that the robot will execute for a specific set of environment conditions, (iii) Why did the robot NOT execute a particular action? The method works as follows. The human user will write their query in natural language. Consider a self-driving vehicle. A user may wish to ask the vehicle, "When do you turn right?" The query is mapped to a template such as "`When (action=right)`" and the method will then analyze its policy in the state space to determine states where the robot is likely to execute this action. There is a state mapping to convert this list of states into natural language, and thus a text explanation can be delivered to the human user. This is achieved by representing world states as the composition of predicates. The authors tested it on gridworld, cartpole, and a simulated robotic parts inspection task. In the latter, the robot must find and analyze physical components that have been placed in front of it by a separate robot agent. In one example, the researchers put one of the parts to be analyzed far away from the robot, which was stationary. The robot did not analyze the part, and the researchers asked the robot, "Why didn't you inspect the part?" The robot replied, "I did not inspect the part because I cannot reach the part. I inspect the part when the stock feed is on and I have detected a part and I can reach the part."

3.11 Model Reconciliation

Model Reconciliation [10] is for a very specific type of explanation, explaining the *difference* between a robot's learned model and a human model, or what the human might expect.

3.11.1 Certain Model Reconciliation

> Post-hoc; Local; Model-specific; When-Produced: After; Explanation Type: text; Has been applied to robotics in real world; Knowledge Limits: Complete; Explanation Accuracy: Certain

3.11 Model Reconciliation

For certain model reconciliation, a complete knowledge model is required, represented in a Planning-Domain Definition Language (PDDL) format [121]. The agent can compare the human model to its own and describe differences, in terms of PDDL elements. For example, if the action is not performed and the human asked why, it might tell the human that the action has some preconditions that are not fulfilled. Korpan and Epstein [152, 153] apply this approach to robot navigation in a task where a robot must search several rooms in order to find the target. The agent can be asked why it did a particular action, why it did not do a particular action, and how sure it is (thus incorporating Knowledge Limits).

For example, at a particular time the robot might be asked, "Why did you do that?" and answer, "Although I don't want to turn towards this wall, I decided to turn right because I want to go somewhere familiar, I want to get close to our target, and I want to follow a familiar route that gets me closer to our target." The agent may be asked how sure the agent is in their decisions. It might say it is sure or it might say, "I'm not sure because my reasons conflict." When asked "Why not do something else?" the robot might answer "I decided not to move far forward because the wall was in the way" or "I thought about shifting left a bit because it would let us get around this wall, but I felt much more strongly about turning right since it lets us go somewhere familiar and get close to our target."

3.11.2 Uncertain Model Reconciliation

> Post-hoc; Local; Model-specific; When-Produced: After; Explanation Type: text; Has been applied to robotics in real world; Explanation Accuracy: Certain

An extension to [152, 153] is undertaken by [50, 285] for the purpose of performing model reconciliation without full knowledge of the human's model. It generates an explanation despite having uncertainty about which of different possible combinations of the human model could be correct. Additionally, these new methods are interactive with the user. The following motivational example is provided by [152]: The design of the Fetch robot "requires it to tuck its arms and lower its torso or crouch before moving–which is not obvious to a human navigating it. This may lead to an un-balanced base and toppling of the robot if the human deems such actions as unnecessary." In other words, the human perception of the move action and the robot's understanding of the actual move action contain different assumptions about the preconditions for that action. Through repeated query and response, these differences can be addressed and the human elucidated.

3.12 Causal Methods

> **Counterfactual Causal SCM**: Post-hoc; Local; Model-specific; When-Produced: During-Byproduct; Explanation Type: text counterfactuals; No, has not been applied to robotics; Explanation Accuracy: Certain
> **Causal Trees/Graphs**: Self-Explainable; Global; Model-specific, When-Produced: After; Explanation Type: tree or graph; Has been applied to robots in simulation; Explanation Accuracy: Certain
> **Causal Influence Models (excluding counterfactuals)**: Self-Explainable; Global; Model-specific, When-Produced:During-Intrinsic; Explanation Type: causal influence model; Has been applied to robots in simulation

Causality [229–231] is the study of why things happen or, more concretely, what events or variables might cause other events or changes in variables to occur. In recent years there has been work on combining causality with reinforcement learning. In this section we look at the small subset of that work that deals with explanations for robotics.

Some causal RL we have explored already, such as the Counterfactual Causal SCMs in Sect. 3.3.2. Also, [267] (see Sect. 3.5.3) infers hidden variables and environments that cause non-robot agents to act contrary to expectations. Simpler causal methods can be used for useful explanations as well.

Similar to the work on causality for counterfactuals, [188, 189] attempt to create action influence diagrams for various simulated environments. These works focus on "opportunity chains," which they describe as "A enables B and B causes C." They learn a causal model that includes opportunity chains and then formulate it as a decision tree. This tree can be easily queried for information about why the agent performed an action or why it didn't, and explanations are provided to the user.

Fadiga et al. [85] introduces a method of learning a causal model from a combination of observations and interventions in an environment. This method considers which variables are susceptible to intervention and which aren't and plans accordingly. They test on a smart home simulation. Different objects and people in the home can affect each other. They manage to create a causal graph that comes close to the ground truth.

In any environment, an agent cannot affect everything. Knowing what it is possible to affect is crucial. Seitzer et al. [264] tackles this challenge, introducing a method for Causal Influence Detection, determining what variables are possible to modify, a question that is situation-dependent. This can involve prerequisites (an agent must make contact with an object to move it) and uses the causal influence model. This also has the potential to speed up training (if an agent knows there is a causal prerequisite for an action it wants to take, it can seek to satisfy that instead of exploring randomly). They suggest incorporating it into the reward, making it intrinsically desirable for an agent to "gain control over its environment."

In the robot manipulation task on which they tested their technique, the agent learned to grasp, lift, and hold an object after only 2000 episodes by emphasizing causally important exploration.

3.13 Reward Decomposition

The fundamental goal of a reinforcement learning agent is to maximize the reward function it is given. Sometimes engineering this reward function can aid in achieving interpretability. A user study has shown that it can help with understanding counterintuitive behavior [15].

3.13.1 Standard Reward Decomposition Methods

> Post-hoc; Local; Model-specific; When-Produced: During-Byproduct; Explanation Type: Numerical weights of reward components; Has been applied to robotics in simulation; Explanation Accuracy: Certain

A good example of a basic way to use the reward function for interpretability is demonstrated by [144], in which the reward function is a sum of components, each of which has semantic meaning, and Q values are tracked per component. These can be summed to choose actions or looked at individually to determine how a component of the reward influences the action choice. This paper specifically utilizes the concept of Minimal Sufficient Explanation (MSX), which is the concept of identifying the most important reward components for a particular decision and presenting them to the user. They determine "most important" by calculating the smallest set of possible reasons (reward components) that would justify choosing the chosen action over a different action. (A similar concept to MSX has been used by other works in the survey as well and other concepts unrelated to reward decomposition.)

This technique is utilized for robot navigation on a turtlebot in [137], where the MSX and a well-crafted reward function allow learning "why" a robot made certain decisions. For example, in the case of many dangerous proximate obstacles, the robot will choose to stop, but if there were an obstacle that it could avoid by slowing down without stopping it would do so.

3.13.2 Model Uncertainty Reward Decomposition

> Post-hoc; Local; Model-specific or Model-agnostic; When-Produced: During-Byproduct or When-Produced:After; Explanation Type: number, numerical weights of reward components, sometimes raw numbers sometimes transformed into words; Has been applied to robotics in simulation and real life

Another type of reward decomposition takes the "Probability of Success" approach, also referred to as "model uncertainty" [124] or a "memory-based" explanation [61, 62].

In Q-learning, during exploitation, an agent will take an action that has the highest Q-value. The value of this Q-value is not particularly communicative. In [62], this is converted into an explanation regarding how probable success is so that the agent will tell the user, "I am choosing to go left because that has a 73.6% chance of reaching the goal successfully." The paper presents three ways of calculating the probability of success. It can be calculated explicitly by storing transitions in memory (of course this limits scalability). More feasibly, it can be estimated by tracking the P-values (probability of success values) in parallel with the Q values, updating them during training based on whether an episode is successful or not. (This method is thus also a type of Meaningful Representation, as in Sect. 3.4.2, produced as a byproduct.) Finally, the relationship between Q-value and reward can be used to calculate what that Q-value means with regards to probability of eventually reaching a future reward. Specifically, if the final reward is R^T and the discount factor is γ, then $Q(s, a) \approx R^T \cdot \gamma^n$, and $n \approx \log\left(\frac{Q(s,a)}{R^T}\right)$, where n is the number of actions it will take to reach the goal. They use this along with assumptions about the stochasticity of the environment to estimate the probably of success. They test this on a simulation of robot navigation.

A user study [63] has demonstrated how Probability of Success explanations can be well understood by non-expert users of AI.

One limitation of this type of reward-derived q-value inference approach is that it requires a sparse reward, or at least one where the goal is a much higher reward than is received in an intermediate manner, and where there are not many different types of intermediate rewards in the environment. It will work for simple tasks with simple rewards.

Reward Augmentation and Repair Through Exploration (RARE) [292] looks at an explanation not in the case of a solitary robot agent but in the context of human-robot collaboration. The reward function is decomposed just as in standard reward decomposition. The goal of RARE is for the robot to infer the humans' implicit reward function, compare it to its own, identify the missing pieces, and communicate this to the human. They test this with a physical Rethink Robotics Sawyer robot playing a game with a human.

3.14 Visualizations

> The attributes of visualizations cannot be generalized.

Visualizations are an important means of communicating information to humans. Creating a truly useful visualization is a challenge since a policy and environment are represented numerically for computation. Many methods in this survey utilize visualization, and in this section, we note a couple that focus on visualization specifically.

One use of visualization is to help with debugging reinforcement learning. The framework outlined by [73] seeks to answer the questions "What sequence of states causes what behavior?"; "Which states lead to similar outcomes?"; and "How does the distribution of states the agent visits change over the course of training?" They show this visualization to the user using a graph with a small number of dimensions and the image with multiple positions of the simulated physical agent superimposed upon itself, as shown in Fig. 3.10. They are also able to identify dimensions of high variance, which could be responsible for a failed training, and allow the user to replay specific trajectories based on specific criteria that users might want to investigate.

There are several other visualization methods discussed in this survey in other sections.

Mishra et al. [197] targets the entire policy, distilling it to a tree that can be used to visualize counterfactuals for some toy RL domains. This is discussed in Sect. 3.3.2 on Counterfactuals. Acharya et al. [2], Huber et al. [133] are observation-based methods with visual components discussed in Sect. 3.5.1. The dimension reduction/abstract MDP methods [341, 356] from Sect. 3.4.1 also contain visualizations, and any method in that section could easily be visualized. Of course, various saliency methods use visualizations: [26, 204, 205, 219, 239, 284, 340] in Sect. 3.2.2, [133, 186, 323] in Sect. 3.2.1, and [89, 104, 111, 132, 336] in Sect. 3.2.3.

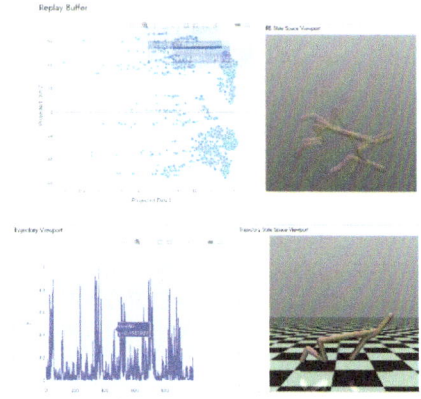

Fig. 3.10 In these images from [73] we see examples of visualizations useful for debugging. At left, a distribution of states across two dimensions and a visualization of those states; at right, a user has selected a specific state from a sequence during training, seeing it visualized along with information about its error

3.15 Instruction Following

> Self-Explainable (Post-hoc for **Human-in-the-loop**); Local; Model-specific; When-Produced:Before; Explanation Type: instructions; Has been applied to robotics in simulation and real life; Knowledge Limits: range from Not Attempted to Partially Attempted

Some robotic systems involve humans giving instructions to a robot, whether learned or preconfigured. These instructions themselves have semantic meaning and can be used to understand what the robot is doing. Consider [93, 94], where a human gives an agent instructions and the agent follows the instructions. Later, the agent can communicate to the human an explanation for its actions by associating its chosen actions with the language the human used to request the specific actions. (These two papers focus on being able to perform this association even as the policy changes, after the initial instructions are given.) In another example, Interactive Transferable Augmented Instruction Graphs were used to teach a robot tasks and new commands composed from smaller known commands, representing a test plan as a graph that is human interpretable [248].

A learning approach on a simulated manipulation task is provided by [268], where DRL is used toward following natural language instruction using a Learning from Demonstration paradigm. Venkatesh et al. [310] builds on [268] in an interesting way. In [310], human experts create Python functions that control robot movement, where each Python function corresponds to a natural language instruction (not dissimilar to Interactive Transferable Augmented Instruction Graphs (ITAIG) [248]). This is the dataset that is used during imitation learning. The reinforcement learning agent learns to predict which Python functions to use (as opposed to which raw movement action to use–the RL action is the choice of Python function), and these functions control the real-world robot for a manipulation task. A further simple real-world task using instruction following is demonstrated in [193].

We also note how instructions can be combined with HRL, as in [142], described in Sect. 3.9.2, where instructions are used to build skills that are then used in hierarchical reinforcement learning, mixing the two kinds of explainability. In [145], the agent imagines a goal based on instructions and then attempts to reach that "imagined" goal. They develop a latent space where the state is represented in a way that refers to the spatial positioning of objects. Then they create a model that maps from human language to this space. This space can be sampled from, or a human language instruction can be converted into this latent space, corresponding to a state that may exist but that the agent has not seen. Then, the agent treats this state as a goal and attempts to reach it. See the diagram in Fig. 3.11.

While [142, 145, 268, 310] may not have had explainability in mind, the relevance is clear. When the robot has learned the policy where the actions correspond to natural language or code functions, there is interpretability built in.

3.15 Instruction Following

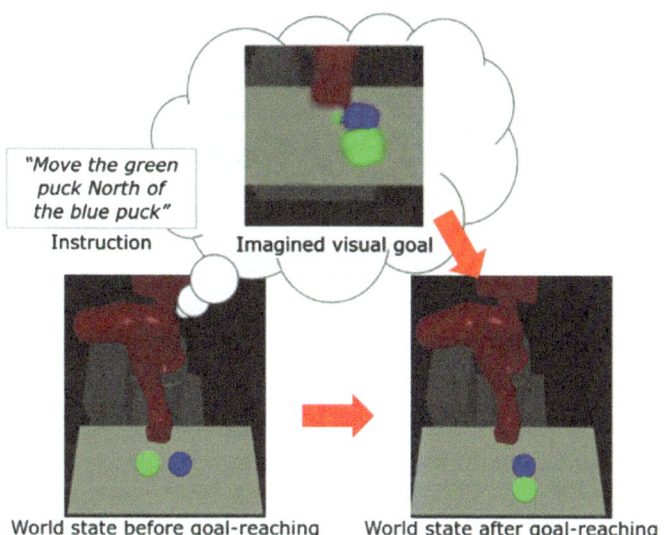

Fig. 3.11 In this image from [145], an instruction is transformed into an "imagined" state, which becomes a goal that the agent attempts to reach. (The "green puck" is the lighter-colored puck, which begins below the darker-colored "blue puck")

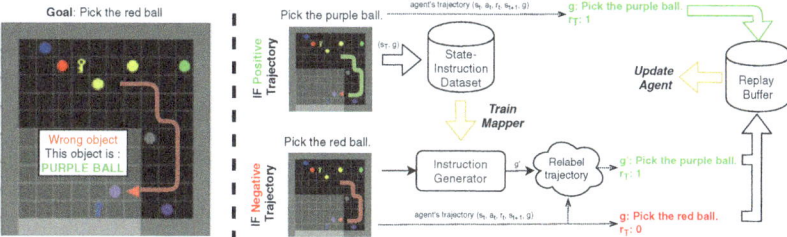

Fig. 3.12 This diagram shows the process from [57]. Correctly followed instructions get added to the replay buffer. Incorrectly followed instructions (i.e., picking the purple ball (at position (9,6)) when asked to pick the red ball (at position (2,3))) get relabeled with what the correct instruction would have been to generate the demonstrated behavior. Although [57] does not make the connection we do in this survey, the use of these instructions and their association with trajectories can help facilitate interpretability of the eventual policy

An alternative approach to instruction following is found in [57], which extends hindsight experience replay [18]. In a situation where certain instructions are followed and some instructions are not followed, the folded trajectories are relabeled with instructions that accurately describe what the robot did. In this manner, the robot can not only learn which instructions it followed correctly and incorrectly, but also how to correctly describe what it did during the "incorrect" trajectories in the instruction-space (i.e., in terms of language instructions). An example is shown in Fig. 3.12.

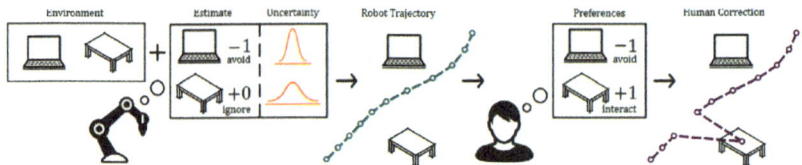

Fig. 3.13 These diagrams are taken from [183], which provides the following description: "Given an environment and the current estimated preferences, the robot selects a trajectory (teal). The human then observes this trajectory and provides a correction to better match their true preferences (purple). Traditionally, the robot uses the correction to update its estimate at the next iteration. We propose that the robot should also obtain the uncertainty over this estimate (orange)"

In the subcategory "Human-in-the-Loop Correction," robots learn from human corrections (which we consider a kind of instruction, relative instead of absolute). In [183], iterative inverse reinforcement learning is used to learn from human feedback, accounting for what the algorithm does not know. They test this on a simulated delivery robot. There is previous work that follows this approach, and this paper improves on the previous state of the art by estimating uncertainty in the agent's current knowledge. Find a diagram in Fig. 3.13.

3.16 Symbolic Methods

Symbolic methods are widely used in classical artificial intelligence [32, 72, 98], and much work has gone into combining them with more contemporary learning methods [46, 99, 139]. In this section we review a few ways symbolic methods can be used or integrated into reinforcement learning in robotics.

3.16.1 Symbolic Transformation

> Post-hoc; Global; Model-agnostic; When-Produced: After; Explanation Type: Symbolic Structure; Has been applied to robotics in simulation and real life; Explanation Accuracy: Certain

We have looked at how a policy can be a decision tree or a neural net. It can also take the form of a symbolic structure. There are multiple ways to achieve this structure. One way is via policy distillation. In [128], each action is a sub-policy learned by a demonstration and human intervention. Each of these policies consists of logical predicates. This method is demonstrated on a robot social navigation task (simulation of a robot in a busy office hallway). The policy actions can include commands such as follow, halt, and pass. Predicates involve

3.16 Symbolic Methods

evaluations of the environment. The learning agent is given a sequence of demonstrations, and it will fit a symbolic policy to it by searching over the predicate space. The authors of this paper demonstrate how the agent can learn to be aggressive/greedy or more subservient/nice in terms of balancing time-to-goal and leeway given to humans. The policy can be adapted if it encounters a change in human preferences by modifying the parameters of the predicates. The policy is interpretable.

Another means of achieving a symbolic policy via distillation is [162], which considers possible symbolic transformations of a policy generated by an autoregressive recurrent neural network. The policies can be represented as mathematical expressions, intended to be more understandable than a neural net.

3.16.2 Symbolic Reward

> Self-Explainable; Global; Model-agnostic; When-Produced: During-Intrinsic; Explanation Type: Symbolic Rewards; Has been applied to robotics in simulation only

One unusual approach found in [269] uses a symbolic approach in the reward function. This paper tackles a problem with a sparse reward and learns symbolic rewards that can be used for intermediate rewards for training. There is a series of symbolic regression trees that map raw state observations to a scalar reward based on the logic of the symbolic model. Each node of such a tree represents a simple mathematical or logical operator sampled from a predefined dictionary of operators or basis functions. In this paper, different trees are grown and selected by an evolutionary fitness algorithm. It's tested in a mujoco environment and the Google Research Football environment [159]. In the case of the sparse reward environment, this intermediate reward can aid in learning.

3.16.3 Symbolic Learning

> Self-Explainable; Either Global or Local; Model-specific; When-Produced: During-Intrinsic or When-Produced:During-Byproduct; Explanation Type: Symbolic Structure, Graph Representation, and where applicable a plan upon the graph; Has been applied to robotics in simulation and real world

Of great interest are methods where we can learn a symbolic structure for a policy directly without distillation. A prime example is [175], which learns and utilizes a symbolic graph representation for the purpose of tackling a manipulation task. A robot manipulator arm is a complex mechanism, and the action and state space for the manipulation problem can be quite large. Conceptually, however, the problem can be simpler if it is thought of in terms of spatial relationships between objects. A graph is one way to give this understanding to the machine. The manipulation task they choose is block stacking. They assume the existence of a parameterized primitive action that can accept an object and a goal, and which will grasp the object and place it in that goal. All objects and possible goals in the scene are represented as nodes in the graph, and all nodes are connected to all other nodes with information including object type, position, and (if a goal location) whether it is empty. While they are not explicit about how they create the graph, they allude to doing something similar to [172, 274]. They use a few expert demonstrations and a neural net to train the policy, which takes this graph as input. They represent it as a probabilistic classification problem over the actions, objects, and goals. The goal of the graph neural network is to predict which object to manipulate towards which goal. They also show how a GNN-Explainer [337] can be used to provide interpretability for their graph neural net policy.

For a hierarchical learning scenario where the high level policy is symbolic, using primitives that operate in a continuous lower level space, the process of abstracting to primitives can cause information loss. Bilevel Neuro-Symbolic Skills is an approach that tries to reduce this information loss [279]. In this method, the primitives are parameterized predicates. The algorithm considers which predicate primitive policies it is possible to execute when it is making its plan.

Elsewhere in this survey we discussed some additional symbolic learning methods that overlap with other categories. The above example is a sort of state transformation, and we find further symbolic learning for state transformation discussed in Sect. 3.4.2, with Meaningful Representation Learning in the form of learning logical relationships [343] and learning a graphic representation of visual input [274]. Symbolic methods are a natural fit for combining with hierarchical methods. In Sect. 3.9.1, find a discussion of using or learning skills as part of the hierarchical symbolic model or policy [96, 334].

3.17 Legibility or Readability

Self-Explainable; Global; Model-agnostic; When-Produced: During-Intrinsic; Explanation Type: the chosen actions themselves are more intent-expressive; Has been applied to robotics in simulation only; Legibility: Attempted

Legibility and readability are robotics concepts that could be considered a subset of explainability. Legibility is a robot choosing actions that allow an observer to correctly infer its goals, and readability is a robot choosing actions that allow an observer to correctly infer its future actions. Examples of legibility outside of an RL framework include [53, 182, 261, 263, 311, 319]. (Find a further description of these concepts in Section 2.3.) In the realm of reinforcement learning as applied to robotics, these concepts remain understudied.

3.17 Legibility or Readability

Another relevant paper that deals with these concepts is [36]. In this paper, the observer gives feedback regarding the observer's estimate about how likely they think it is that an agent is going for a particular goal. This feedback is a signal that can be used to improve the robot's legibility. They demonstrate that trajectories become more legible when this feedback is integrated.

Legibility can be incentivized through the reward function. For example, the reward function for a simulated robot arm grasping task in [352] includes a term that incentivizes the path to cause the end-effector to have a small Euclidean distance to the end goal (among other factors), instead of what might be a more optimal path from a motion-planning perspective, but which would rotate the end-effector in a manner causing lack of goal-comprehension/prediction on the part of a human observer. A user study demonstrated the effectiveness of this approach for increasing legibility.

Regularization for Explainability, from [232], adds legibility to RL using techniques that come from the sub-field of explainable planning. They give a formal description of legibility, "The main goal of [legibility] is to bring the intention predicted by the observer's model close to the intention of the agent, and to maintain such closeness in time." A corollary to that is that a legible policy is, as a prerequisite, discernable from other policies; that is, an observer can tell different policies apart to some degree while watching the behavior of the robot. The method in [232] takes this into account and attempts to select actions that maximize the chance of an observer correctly distinguishing the currently selected policy from alternative policies. They test this with a task where a simulated agent must navigate through a tunnel with multiple-colored squares, travelling only on squares of particular colors while avoiding squares of certain colors. The choice of which squares to traverse can help indicate to a human observer which colors fall into which category (in comparison to a valid tunnel traversal policy which causes a human observer to infer the status of some colors but remain uncertain as to whether one or more other colors is/are allowed or proscribed).

A framework for evaluating legibility in [320] further notes the work remaining in bringing legibility to RL methods.

4 Key Considerations and Resources

In this chapter we offer additional discussion of the techniques described in the preceding chapters. We raise additional issues, provide insights and describe trends, and note additional resources relevant for practitioners.

4.1 General Discussion

Multiple types of explanations may be necessary to ensure that a human user understands an agent's mental model. A 124-person user study provided groups of humans with saliency map and reward decomposition explanations for a real-time strategy game, learned with reinforcement learning. With four groups for neither explanation, one of the two explanations, or both explanations, only the group with both explanations showed a statistically significant improvement in helping a human build a mental model of the agent [15].

Another key consideration when discussing explainability is the purpose towards which an explanation can be put. This is related to the concepts of Audience, Format, and other attributes. In addition to innovation in the method of explanation, it could be beneficial for community members to share examples of how explanations can be used in various applications. For example, [125] uses explanations to defend against an adversarial attack on a UAV.

There are some types of explainable methods that could be applied to robotics without reinforcement learning, where using reinforcement learning nevertheless adds benefit. When considering causal methods (Sect. 3.12) for example, there is evidence [181] that "experience" (i.e., interacting with an environment as occurs during RL) is important for determining what is causally or acausality linked. We are thus seeing the beginnings of combining causality with RL. We provide some specific suggestions in the "Causal" paragraph of Chap. 5.

A key challenge when using learning methods for robotics is processing the massive amounts of high-dimensional data the robot may have as inputs and determining how to transform them into more useful data. Many methods from our survey seek to deal with this, including Dimension Reduction (Sect. 3.4.1), Meaningful Representation Learning (Sect. 3.4.2), and Symbolic Learning (Section 3.16.3). Simple dimension reduction is useful, and creating a meaningful space is even better. Oftentimes, if we want to end up with some structure, we must create that structure ahead of time. Many researchers are investigating how to generate usable structures from raw pixels or data, such as [67, 83]. Generating interpretable structures is of particular interest in Human Robot Interaction (HRI) applications, as well as domain transfer and Simulation-to-Reality (sim2real) [214, 339]. Understanding why some skill does or does not transfer would be valuable. Additionally, the process by which this interpretability is introduced (e.g., if a hierarchical method is used, as in Section 3.9), could help with domain transfer itself. Furthermore, as shown by [327], learned representations in different tasks in similar environments can be similar and used for task transfer. Similarly, structure can be used to diagnose faults in the sim2real process.

To further underscore the value of Dimension Reduction and similar goals, a human study [287] showed with a simulated mars rover that natural language explanations based on a smaller number of features (excluding irrelevant features) is more comprehensible to humans than explanations that try to account for all the features of a space. This is also useful because it implies to a human which features are important via the explanation. In a related vein, [342] explores and argues for calibrating the level of abstraction of agent-generated explanations to account for the human's "cognitive load." This might be challenging to estimate but is an interesting and useful goal.

At the same time, it is interesting to note that in [287], while the more "focused" explanations were more understandable, *preference did not always match the more focused, understandable explanations*. Of the 18 subjects, 8 preferred longer explanations and 8 preferred shorter explanations. Some liked conciseness and explanations which highlighted importance, while others appreciated the extra information. So, preference for complete versus focused explanations may be a behavioral trait. This is something future work could explore that is relevant to XRL-Robotics. There was also a half and half split for preference between bar graph explanations and natural language explanations. Future work could pursue customizable explanations, perhaps even creating a mechanism to personalize explanations, modifying the level of explanation or type of explanation based on personal preferences. Either of these suggestions could be combined with any of the above methods or the future work suggestions below.

4.2 Limitations of Some Methods in this Survey

In this section, we discuss of additional limitations of some of the method categories previously covered in this survey.

One of the dangers of relying on saliency maps is that it can be difficult to ascertain the degree of explanation accuracy. Since the space consists of many pixels, it is very possible that some are given credit when they are extraneous/less salient. It can be difficult to falsify a saliency method and determine if it is correct or not. There is no ground truth since it is often attempting to explain a black box method; indeed, that is why it has value. Because of this, [24] suggests that saliency maps be considered exploratory and not explanatory. While it is important for a human user to keep in mind these very real limitations of the method, there is also work related to ensuring that the allegedly salient features are in fact salient [111]. More work could be done in this vein.

One path towards addressing this could be to compare machine attention to human attention. A user study in [109, 349] compared what features an RL agent paid attention to compared to what features a human paid attention to as they performed the same task. They determined that some of the policy errors in the RL agent were because it failed to pay attention to that to which the human was correctly paying attention. This information could be useful for debugging an agent and for improving saliency methods themselves.

Observation-based methods (Sect. 3.5.3) are limited by assumptions humans make. For example, one might expect that measuring the most frequently visited states is informative, but such states may not be the most important for achieving the expected reward [236]. This is an important consideration for designers of such methods.

Another insight from [236] points out that RL policies are sometimes more complex than is truly required to solve a task, which increases the challenge of analyzing such policies. This has ramifications for observation-based methods (Sect. 3.5.3), Causal Methods (Sect. 3.12), and Counterfactuals (Sect. 3.3). Typically, an RL policy makes decisions every time step, but sometimes in the non-critical states, the choice of action is not so important, and a default action would suffice. Thus, a simpler policy than the one achieved often exists to solve a task. This is also implied by the work on network distillation [256]. This applies to both white box and black box methods. Outside the scope of this survey are methods for simplifying black box policies, although [236] offers one such solution based on identifying which decisions are important and which are not.

A recent user study demonstrates how certain policy summaries can cause harm to a human [7]. The example they have is simple: in a simple strategy game (similar to one called "Noughts and Crosses"), literally correct text explanations can cause a human's performance to become worse under some conditions if the explanation is too complex to understand. This result should give researchers pause. Investigating this on more complex and realistic environments and use cases is also a worthwhile future direction.

The use of VAEs [147, 244] (see also Sect. 3.4.2) is growing rapidly. However, [52] argues that if they are used on images and if the images contain too much information irrelevant to the state, it won't work as well. They demonstrate this by creating a siamese network with a contrastive loss and applying it and a VAE approach to a block stacking task. They find a more useful representation with the siamese network than with the autoencoder approach.

Finally, XRL for robots has all the limitations present in standard robotics. The dangers and demands of physical hardware are often not considered in RL work that does not involve a physical machine and must be considered by the researcher or practitioner who intends to do so. A recent study [135] took an impressive earlier method (related to autonomous robot arm grasping [168]) and found that the best policy learned was "sensitive to a hardware degradation of the fingers, which caused a consistent performance drop of 5% in as little as 800 grasps executed on a single robot." They suggested mitigating this using A/B testing, as in [296]. Regardless of the approach, this and other concerns of wear and tear could be considered in explanatory contexts as well. A researcher could consider incorporating explanations that detect and report on wear to a machine. Related to Knowledge Limits and Explanation Accuracy, it is worth considering that if some explanatory model is developed to explain a robot to a human, in some circumstances the quality of this model may degrade as the robot does (if it stays constant as the robot changes).

4.3 Human-Robot Interaction Considerations

Every interpretability and explanation method implicitly assumes the existence of a human agent, the needs of which the method considers, whether explicitly or not. Thus, there are further insights to be gained from the field of Human-Robot Interaction.

Many have explored ways robots can give explanations to humans [38] or otherwise influence humans [251], irrespective of reinforcement learning. While some methods are not relevant or applicable to reinforcement learning, there are plenty of methods that are compatible despite not targeting integration with RL directly. Any problem that can be expressed as an MDP can take advantage of, for example, the work of [38], and some problems RL solves can be expressed as an MDP.

We also refer the reader to [115], which studies *how* a user might want a robot to explain itself. While some developers create explainable methods based on what is possible, it's important to always keep the user in mind and, as a starting point, ask what is most useful to users. Beyond user preferences, we can study the psychology of end-users, as [324] does, to learn how humans trust robots and form their mental models, so as to better craft helpful explanations. As [146] notes, explanations do not automatically increase trust. One cannot simply assume trust increases because an explanation is present; the explanation must be appropriate to the task and the human.

Bridging the robot-human communication gap is a common goal. Sumers et al. [289] attempts to formalize the ways in which human language expressions map to RL concepts. At a high level, for example, they posit that "instructions" communicate requirements of the policy and "descriptions" are relevant to shaping the reward function. Further work could

4.3 Human-Robot Interaction Considerations

be done exploring this or other attempts to map human language specifically onto an RL paradigm for robots.

Another factor to consider is that different populations may have different needs. In Sect. 2.3, we discussed the Audience attribute as a means of comparing different approaches.

One axis to consider when thinking about the Audience is human role. An end-user using a finished product, an engineer implanting learning algorithms, and a researcher developing new learning approaches all will want and need something different from an explanation. An end user requires simplicity in their explanations, while a researcher or engineer might want more detail. An engineer cares a lot about diagnostics and debugging. A focus on end-users entails considering trust and adoption. A researcher is looking to scientifically validate new approaches. We recommend that any algorithm creator or user think concretely about who from this list they have in mind when they are building their method or system.

In addition to the type of user, within the end-user population, there is a wide range of demographic and other variation. For example, the age of the user is also a consideration. Seven-year-old children can benefit from explanations, as shown by [304], which demonstrates how a robot-child team can improve at the performance of a task as a result of contrastive explanations. Although not discussed explicitly below, further work regarding real-world users of different populations and demographic groups distinct from the researcher's groups would be valuable.

All audiences benefit from simpler explanations. The previously referenced [206] from Sect. 3.3.2 presents a framework and user study for using contrastive explanations to aid human understanding. Another user study, in [276], investigates tailoring robot explanations to specific humans through meta-learning.

Most papers surveyed here involve enabling a human to understand a robot, but there is also much work on a robot understanding a human. In the "Human-Interpretable Theory of Mind" [221], an agent develops a theory of mind of a human and explanations for a human. This paper is neither reinforcement learning nor robotics, but the authors of [221] performed a human-AI teaming task in Minecraft and discovered that the explanations and theory of mind approach improved predictive accuracy as well. The paper itself notes these ideas could be extended to RL, and this could be done in a more realistic reinforcement learning environment for robotic search and rescue.

In general, it is instructive to consider whether any systems we think of as purely autonomous should be recast as a human-robot team. For example, [293] discusses how explanations aid in human-AI-robot teaming. Specifically, in the application of autonomous driving, it helps with the transfers from human to vehicle, when such is required. Much work has been done on autonomous driving in the reinforcement learning community, although not as much has been done on explainable driving, and it would be useful for the XRL community to consider framing autonomous driving as human-robot teaming and not simply an autonomous robot carrying a passenger. Additionally, [293] does not use RL, and integrating the idea of safe driving hand-off with an existing RL approach could be feasible future work.

4.4 Legibility and Readability

Most methods do not involve legibility or readability. We want to call out a few methods to supplement the discussion in Sect. 3.17.

From Sect. 3.13.2, RARE [292] is a reward decomposition method that estimates a human's implicit reward function and helps the human understand how it differs from its own understanding of reward. This could be considered a sort of readability.

In Sect. 3.2.2, one paper puts a Pepper robot in a public space for 14 days for human social interaction [239]. There are elements of legibility and readability in the way that it responds to the humans to convey its intent or desire to enter a discussion.

Still, in both cases, readability and legibility are not thought of using this terminology or explicitly from this perspective. This is a widely studied topic in robotics, and there is a great opportunity here for valuable work. It is surprising how little has been done. Further work into robot legibility and readability in the context of RL methods is important.

4.5 AI Safety

A field that overlaps with the topics covered by this survey is "AI Safety," which is concerned with making safe AI [56, 60, 105, 123, 224]. This field covers both "short-term" safety, which refers to present and near-term risk (for example, a self-driving car injuring a pedestrian) and "long-term" safety, which seeks to prevent harm coming from AI systems that will be developed in the future. (An example of a long-term risk might be a sophisticated hospital patient care system with the goal of maximizing patient happiness. Improperly set up, one can imagine the system choosing to let unhappy patients die to increase average happiness rather than pursuing a course of life-preservation as a human doctor would. Naturally, if the outcome in this simple example is foreseen, it can be prevented, and in a system today it could be noticed. However, without foresight and planning, a future system may be so complex and opaque that more subtle but no less dangerous errors go unnoticed.) It is relevant to reinforcement learning because a big concern in the field is correctly conveying human desires to the machine (since incorrectly understood desires could result in harm). This is also a topic of interest in reinforcement learning, which has whole subfields related to reward shaping [163, 171, 329] and inverse reinforcement learning [22, 215] to tackle the same problem. It is of relevance to our interest in robotics, because an embodied agent can have a greater effect on the world, and thus has larger safety concerns.

Another concept from AI Safety is "value alignment," that is, the degree to which a robot's goal is the same as the goal a human would want it to have. After all, human goals can be difficult to express and encode, and sometimes human desires are difficult for a human to understand. Sanneman and Shah [260] is relevant to the topics covered here, as it discusses which types of explanations are the most useful for value alignment.

Explainable and interpretable systems contribute to AI Safety. Some methods previously covered in this survey included Safety as a goal:

- NLP Templates for converting models or policies to text [190, 322]
- Decision Tree Methods:
  - Conservative Q-Improvement [252]
  - Mixture of Decision Trees by Distillation [309]
  - Decision Tree Policy Learning and Modification [249, 250]

- Saliency Visualizations (by backpropagation) [133, 186]
- Symbolic Model or Policy [96, 334]
- Safety-Constrained Learning [16, 299, 335, 345]
- Safety-Informed Execution [155, 187, 227, 294]
- Safety in Execution: Safety via Planning [82]
- Safe RL via Counterexamples [95]

There is also interest in the AI Safety community for safe "superhuman AI." Since this superhuman AI might include robotic systems, we note it here. Andrulis et al. [17] discusses challenges involved in creating interpretable/trustable superhuman XAI and proposes what might be needed for it. An unsafe superhuman General Artificial Intelligence [30, 208] can be referred to as "existential risk," (or X-Risk) since it could do harm on a massive scale [56, 60, 224].

4.6 Environments

There are several simulation environments for conducting reinforcement learning experiments on robotics that may be of particular interest to those pursing explainable or interpretable methods:

- **S-RL toolbox** [240]: a collection of environments for testing reinforcement learning state representation methods
- **SafeRL-Kit** [347]: an environment for benchmarking autonomous driving methods from a safety perspective
- **VRKitchen** [97]: a virtual reality-enabled environment where robots can interact with kitchen objects and participating humans
- **Meta-world** [338]: a benchmark and evaluation for multi-task and meta reinforcement learning
- **Alchemy** [321]: a system for testing the ability of agents to plan and reason using latent state spaces and structure, making it relevant to many XRL methods

- **Causal World** [6]: an environment involving robotic manipulation meant for testing methods related to causal learning and transfer learning

We also bring up the "Watch and Help" (WAH) environment, [237], a task for measuring social intelligence in agents. In WAH, a robot helps a human complete a complex household task. The robot will attempt to i) determine the goal by viewing a single demonstration and ii) work together with the human to complete the task. This could be useful for testing legibility/readability as well as other aspects of a robot signaling a human.

While not specific to explainability, HIPPO Gym [297] aims to be an AI Gym-like interface for human demonstrations for RL and other human-robot interaction in simulation.

Also of interest could be [272], a benchmark towards allowing an agent to construct a human theory of mind, which could allow it to make better explanations (related to [221]).

5 Opportunities, Challenges, and Future Directions

One of the contributions of this survey is many suggestions for future research directions. We have already mentioned some suggestions, and the remainder are proposed in this section, grouped by category or sub-category.

5.1 Opportunities, Challenges, and Future Directions

Decision Trees:

Methods that use reinforcement learning to learn a decision tree by an additive process [65, 248, 252] have not yet been applied to complex, real-world problems. This would be worthwhile, since decision trees are useful end-formats, and it could be helpful to be able to create them without an intermediary black box stage. Related to this idea, differential decision trees [277] have been used for reinforcement learning policies but have not been extensively used for robotics. Attempting to apply them to more robotics applications would be worthwhile. If a problem can be expressed as an MDP, it can be wrapped in an Iterative Bounding MDP [301], which facilitates producing a decision tree solution while solving the MDP. This has not yet been applied to robotics.

We have seen trees with some nonstandard structures such as sigmoids instead of Boolean conditions [74, 75] and networks in branch nodes [58]. This idea could be taken in many directions. Consider creating a tree that has a neural net on the leaf nodes for fine tuning. This could be used for low-level control or to allow the use of a tree for continuous action space. In general, it will be worth thinking about mixing in other black box components to nodes of the tree to retain high-level interpretability while increasing performance and the ability to tackle complex problems by using a black box method in the lower level or otherwise mixed in.

Although a single-tree approach is always the most interpretable, when facing a problem that cannot be tackled with a single tree, one can attempt to tackle it with a mixture of trees [44, 309] and focus on analyzing the mixture for the purposes of safety. Applying this to a real robot would also be novel.

Increasing sample efficiency in policy distillation is worthwhile. Unlike standalone RL, a distillation method that processes samples drawn from an existing policy needs to store all those samples and process them, which takes time and memory. There is work towards identifying the most important subset of samples to use in a distillation process [180]. This would be relevant for combining with decision tree distillation methods (e.g., [58, 250, 252] and others in Sects. 3.1.1 and 3.1.2) or symbolic policy by distillation methods (e.g. [128, 162] from Sect. 3.16.1).

Saliency Maps:

Questions have been raised about the usefulness of saliency maps (Sect. 3.2). They seem intuitively interesting, but do they provide actionable information? Do they increase trust? Some studies have investigated this [133], and further user studies along this line may be merited.

In general, user studies to investigate how effective various types of explanations are in real life and how well humans from various audiences understand them are worthwhile. Such studies could be undertaken with regards to any and all of the methods described in this survey.

There are a variety of methods for producing saliency maps that have been applied to Atari 2600 games solved by DRL, deriving saliency information from the values of weights or other information in the neurons or networks [204, 205, 219, 284, 340]. These techniques have mostly yet to be applied to robotics in simulation or real life. Any robotic application that is primarily camera-driven and is addressed using DQN, A2C, or another compatible DRL technique could take advantage of the techniques discussed in the cited papers.

In [239] we see a humanoid-like robot learning to interact with humans over the course of 14 days of real-life learning. Saliency information (i.e., focusing on a human in the view of the camera) is used to help the robot achieve its goals. Future work can do this in concert with providing explanations suitable for human consumption.

Although input perturbation-based methods for generating saliency maps have been applied to robotics [92, 218], more recent work on input perturbation saliency maps (tested on Atari video games) [111] has not yet been applied to robotics. Ivanovs et al. [138] is a survey of input perturbation methods for explaining deep neural networks. The methods benchmarked in [132] are also worth applying to robotic applications.

There are additional types of saliency map methods that were not described in Sect. 3.2 because they have not been applied to robotics yet, including Saliency via Latent Spaces [21] and Saliency via an Attention Mask [270]. Both could be applicable to robots.

5.1 Opportunities, Challenges, and Future Directions

As discussed in Sect. 4.2, many saliency methods do not have the means of verifying that the identified salient features are relevant (as opposed to coincidentally correlated or otherwise incorrectly identified) [24]. Focusing on improving this aspect of saliency would be particularly useful to robotics since robots often have high dimensional input.

One way of achieving this, inspired by [109, 349], could be to incorporate humans in the loop during training. The humans could give feedback not just on the actions but on the attention. This could be explicit feedback but does not have to be. A computer can identify when a human is naturally paying attention to something significantly different from the robot and identify that state or situation as requiring correction in terms of attention, action, or both. If a robot better understands what merits attention by learning from a human, it could speed up overall learning and ensure that the saliency explanations produced are an accurate explanation as well as helpful to both human and robot. Instead of incorporating it as feedback, the human attention information could be pre-collected and provided as prior knowledge as well.

Counterexamples: Methods Useful for Robotics:

Input perturbation for the purpose of creating counterexample explanations has not yet been applied to robotics in simulation or in real life. This could be done using a GAN as in [222, 223] for simulation. Ivanovs et al. [138] is a survey of input perturbation methods for explaining deep neural networks. If semantic features are manually created, the approach of [176] could be followed. In some domains, it may be possible to create these features manually. Otherwise, other works in this survey have described means of creating meaningful features, which could be combined with the use of these generalized value functions to create contrastive explanations.

Causal Counterfactuals:

Pitis et al. [234] uses reinforcement learning and an SCM to improve performance on a robotics control task by using the SCM to generate counterfactual samples that are used as additional training examples. This could be extended to provide explanations for humans using the same SCM, giving explanations of the model itself, giving explanations of why and how the model was improved, or even suggesting future new experiences.

The counterfactual explanations used for the StarCraft agent in [189] could be applied to many robotics applications. In [189], however, they "assume that a [Directed Acyclic Graph] specifying causal direction between variables is given," and they seek to learn the remainder of the SCM and use it. This is a significant limitation. Another important future direction is thus developing methods for discovering and learning an SCM or other causal structure directly from observation and experience during normal training, especially in cases where environment dynamics are not known.

Model-checking Counterfactuals:

The approach of analyzing an MDP to give contrastive explanations, as in [54], is worth pursuing in more complex simulated and real-world robotic environments. One issue with the suggestion is that more complex environments might not have explicit MDPs. Thus, this approach could be combined with some of the methods that attempt to generate an MDP from an environment or generate a high-level or abstract MDP to represent the higher level of a hierarchical process for operating in the environment, such as [25, 302, 341, 356].

Policy explainer [197] is a visualization for counterfactual explanations for a decision tree policy. So far it has only been used in a very simple gridworld domain. This makes sense since it requires a decision tree policy and those are difficult to create for more complex domains. Our survey, however, describes several methods for creating decision tree policies in the context of reinforcement learning. A future direction could be to take one such method, such as CQI [252] or MSVIPER [249, 250], and combine it with policy explainer. Implicitly, this works better with semantically meaningful state spaces. This could be hand engineered or combined with one of the methods from the survey that can create such spaces, such as [175, 274].

State Transformation: Dimensions Reduction or State Aggregation:

Clustering and aggregating state spaces, as in [8, 9], for the purpose of dimension reduction has been done for robots in simulation but not real life. Performing these kinds of techniques in more complex simulations or the real world would be worth doing. Clustering states might seem inefficient given the massive state space of many real-world applications, but if it could be done successfully, the benefits of dimension reduction could be worthwhile for interpretability. (Although it is also worth noting the risk inherent in any dimension reduction technique that the new dimensions–while being some appropriate function of the raw dimensions–might have no human-interpretable meaning.)

Techniques for creating abstract MDPs [25, 302, 341, 356] have not yet been applied to real robots, just in simulation. This is a promising future direction. Additionally, in the case of robotics, an extension could be created that would focus on creating a task plan. For example, one might combine with TAIGs in [248] in some manner to attempt to learn a sequential task plan abstracted from a high-dimensional environment using one of these abstraction techniques.

Meaningful Representations

Consider combining Visual Entity Graphs [274] with any explainable policy technique. Visual Entity Graphs is a technique that improves the interpretability of the state space itself, which is very desirable for combining with almost all the other categories of methods described in this survey. Additionally, further work could be done simply using the interpretable state-VEGs with standard black box techniques. The cited paper uses simple real-world robotics tasks, and this could be extended to more complex tasks. A further challenge would be to extend this kind of work to a mobile robot or other scenarios where the objects may not be as persistent within a scene. (For example, objects and humans a mobile robot will see will enter and exit the frame as it moves, or they will even leave the vicinity entirely.) Creating a graph structure and system that can handle this variability is more challenging than creating one for a scene in which all the entities are persistent and

5.1 Opportunities, Challenges, and Future Directions

only the relationships change. This kind of extension would increase the probability of the interpretable method. Finally, another future direction would be to do this kind of work without requiring human demonstrations. Robots should be able to use a visual entity graph to solve tasks more efficiently without further human input beyond that required to train the graph transformer itself.

One of the limitations of [40, 41] is that they require using reward functions that are compatible with the assumptions they make about "reward proportionality." It could be interesting to attempt to create a similar method that worked for arbitrary reward functions.

The use of human sourced metadata to enable state representations for multitask reinforcement learning is crucial to the success of the method in [282]. Investigating ways that more of the human sourced information could be autonomously determined would be a great step forward.

A VAE is used to create a meaningful latent space that can be used to achieve guarantees in [90]. The learned representation is subject to linear dynamics and thus can be analyzed and the policy trusted. The environment is an Atari racing game. Applying this to more complex robotics tasks would be very beneficial, since the lack of guarantees in learned algorithms is one of the prime barriers to adoption. In particular, if it could be scaled up to situations such as autonomous vehicles, this would have big implications for autonomous vehicle safety and in many other robotic domains where RL is successful empirically, but guarantees cannot be achieved. Similarly, [140] learns the latent space that was compatible with classic PID techniques. They applied it to a robot arm, and future work could be applied to more complex robotic environments.

In the paragraph on saliency maps, we discussed taking inspiration from [109, 349] to compare human and robot attention to improve saliency explanations (and policy performance). A similar approach could be taken in any scenario where human-robot impulses can be compared. In the case of a meaningful representation that is human and robot understandable (such as a graph or tree), we could ask a robot and a human to create an explanation or policy and then compare them. This comparison could be used to improve both the robot policy and the explanation of it.

As noted above, [52] demonstrates a limitation of VAEs for visual state spaces, namely that if images contain too much information irrelevant to the state, it will not work as well. A future work could address this by creating an intelligent VAE with some discriminator or forgetfulness module so that it learns to only focus on important, *relevant* parts of the image.

Observation Analysis:

Amir and Amir [11], Huang et al. [130], Sequeira and Gervasio [265], Sequeira et al. [266] identify critical trajectories that are of great relevance to whether a policy succeeds or fails. They take video of those trajectories to give the user a sense of what the policy is choosing to do and why. This is very easily transferable to robotic domains. Robots are physical agents, and visual examples of robot trajectories can be a layperson-friendly way to convey a general impression of a policy to a human. A way to take this even further would be to combine this with the ideas of legibility and readability and seek to either highlight trajectories that are legible/readable or incentivize the robot to be legible and use this kind of technique to monitor it. Taken in another direction, if there is work done on identifying legible trajectories, this could also be used to identify when the robot is not behaving in a readable/legible manner, which is feedback that might be useful for the engineer developing the robot.

A conceptual framework for strategy summarization is provided by [13]. It involves identifying states of interest to a human and human involvement in the summarization process. They provide detailed suggestions as well as discuss the applicability to a search and rescue scenario.

Safety is mentioned in [131], where it is used as an example of an attribute that could be communicated by highlighted trajectories in the context of autonomous driving. A future direction would be to make safety even more of a focus. Understanding what kind of reward function results in safer versus more efficient driving could be parlayed into giving the user the ability to dial up or dial down safety. Another direction would be to use some derivative of this technique to create a tool to analyze the safety of policies that might not explicitly have safety in mind. All autonomous cars must at least implicitly consider safety, or else they could not successfully drive. However, many may not separate it from overall performance. These techniques could attempt to model the degree to which a safety constraint is satisfied by constructing a reward function with the safety term and performance terms and measuring how it is weighted. A tool could be created to give a sense of the safety to non-technical regulators who could then suggest technical investigation of policies that have worrying unsafe-trajectory-highlights.

In [267], an environment is modeled and, when it behaves contrary to prediction, it is assumed that hidden variables are responsible. The method attempts to identify those hidden variables and the causal relationships with observed outcomes. The application in [267] is a ridesharing recommendation algorithm, but there are numerous additional robotics domains where it is desirable to learn previously unknown causative variables.

The idea of hidden variables is also relevant to the idea of Knowledge Limits. Even if the hidden variables cannot be determined, inferring the existence of such variables can help clarify the degree to which a policy is applicable or accurate. The work of [267] could be extended further in this direction. Knowledge limits is a somewhat unexplored area, and anything that can be done to formalize the measurement of knowledge limits is useful.

Another worthwhile project would be to take the techniques in [236] (discussed in Section 4.2) for identifying important states and applying these techniques to RL for robotics, giving the human user useful information, or constraining one of the explanation methods in this survey to only show the most relevant information. It would be particularly interesting to combine with [265, 266] and other observation-based methods (Sect. 3.5), as well as other methods where examples are used, such as Highlights [11] and counterfactual methods (Sect. 3.3).

5.1 Opportunities, Challenges, and Future Directions

Custom Domain Language:

We noted how PIRL [312, 313] can produce a policy in the format of a domain-specific programming language. So far, this has only been applied to a simulation. It would be worth applying it to a real-world robotics task.

Rudin et al. [254] notes that PIRL demonstrates a drop in performance compared to the black box policy and suggests a challenge of coming up with a method to estimate, for a given black box model, the degree to which an interpretable imitation/version exists that performs above a threshold.

One of the benefits of PIRL is its verifiability. This has been explored somewhat, but future work could focus on it more concretely, particularly with an eye towards safety-related verification, whether in simulation or in the real world. Another benefit to this intrinsically interpretable policy structure is that it is theoretically modifiable and customizable. Just as [250] and [249] involved decision tree policy modification, a future direction would be to allow human or autonomous modification of the PIRL policy. This could be done to improve performance, or it could be done to customize for different user groups or individuals. Consider a home robot or other end consumer robot where an optimal policy may be highly subjective. An end-user may wish for the robot to learn its preferences, or they may wish to program them explicitly. Such a system could build on PIRL to achieve this.

A further application of the modification opportunity is in the realm of human-AI teaming. There are tasks that require a combination of human and AI knowledge or human and AI actions. Having an interpretable policy could be leveraged to facilitate working together, since the human can predict what the robot will do. Additionally, the policy could be modified by the human or autonomously to adapt to the specific team or task situation at hand.

Finally, we note that in this survey there are several different methods for interpretable policies [248–250, 312, 313] and interpretable state representations [40, 41, 175, 250, 274, 282]. Several future directions could involve combining these methods to get higher direct interpretability in both state and policy.

Constrained Learning:

We discussed what we term Safety-Constrained Learning, where safety rules are used to restrict actions or updates during the learning process. Anderson et al. [16], Zhang et al. [345]. This has clear value for a robotics task, but has not yet been applied to real-world robotics. In particular, we suggest using these methods to attempt to train a robot in a real-world task that includes safety concerns.

Constrained Execution:

Some of the safety constrained execution methods [187, 294] estimate an agent's certainty in its own predictions. These knowledge limits are used directly in the cited papers, but the presence of knowledge limits could enhance many of the other types of explainable methods showcased in the survey, making this a direction for future work. Additionally, while some of the planning-based constraints have been applied to real robot, the uncertainty-based constraints have not been applied to real robots. This is future work even without combining with other methods.

Of interest to future work on any safety-informed execution [155, 187, 227, 294] will be the additional ways of modeling model uncertainty found in [1]. Integrating the methods found in the latter into the constrained execution methods discussed above is another future direction.

Hierarchical Skills or Goals:

We shared some high-level interpretability results from hierarchical reinforcement learning [34, 191, 273]. In addition to interpretability benefits, [120] shows how this type of reinforcement learning can be useful for transfer learning. It could be useful to take an existing paper on transfer learning with traditional deep reinforcement learning and apply some of these hierarchical approaches. In addition to gains that could result from using this method in and of itself, having an interpretable structure opens the opportunity to analyzing a new task and modifying a policy ahead of time, using logic, prior to new training. This would be an interesting approach, particularly with the growing interest in policy modification without retraining [249, 250].

Hierarchical Reinforcement Learning with symbolic planning on the higher level and model-free learning on the lower level has been applied to robotics simulations such as self-driving cars [96]. This approach has not yet been applied to real robots, and this represents a future direction.

Primitive or Skill Generation:

There has been much recent work on primitive or skill generation [84, 346], including in the absence of a reward function. However, most of the skills generated lack semantic meaning. A useful future direction would be a project focused on developing a diverse set of skills with semantic meaning.

Template-based NLP Explanations:

Wang et al. [322] demonstrates using a template to convert from a POMDP and trained policy into natural language explanations. Furthermore, it includes confidence intervals in the context of a robot participating in a human-robot team. Wang et al. [322] performs this work in simulation, and extending into reality would be worthwhile.

Maes et al. [190] uses an algebraic formulated human readable language, which is an interesting approach that could bear more exploration. There is room to explore because the cited paper only tested this approach on environments with very small state spaces and action space sizes (small number of actions). Such an approach could be useful for robotics, but only if it is demonstrated on more complex environments. Similarly, constraining a policy to use natural language operators and searching through that space using genetic algorithms, as in [119], is promising but has not yet been applied to complex robotics problems. (However, genetic algorithms have been used for non-explainable robotics, as in the car controller found in [315].)

5.1 Opportunities, Challenges, and Future Directions 85

In [116], a robot learns how to solve a task and can respond to human language queries about his policy with human language responses. This is achieved by creating composable predicates that represent environment states. The use of these predicates as states facilitates reasoning, which allows responding to the queries (after translating between the queries and the predicates using a template). In the paper, these predicates were partially specified ahead of time and partially learned. A good research question would be to ask if these predicates could be fully learned. The challenge here would be mapping from a raw state space to semantically meaningful variables picking out which ones were important. Additionally, this type of approach has not yet been demonstrated on a real-world robot, representing another option for future work.

Model Reconciliation:

Certain Model Reconciliation [152, 153] incorporates Knowledge Limits, but Uncertain Model Reconciliation [50, 285] does not. A future direction could be to incorporate the concept of an agent's decision certainty and uncertainty into Uncertain Model Reconciliation.

Causal:

The method used to find the causal graph for the smart home simulation [85] determined the graph but did not focus on how this would be communicated to the user. This provides a future direction that could be achieved in several ways. One way might be to make use of an NLP template (i.e., [116, 119, 190, 322]), since presumably the graph would be known. Additionally, it could be useful to attempt to apply this kind of technique to environments with mobile agents that are less integrated into their environments. This would present additional worthwhile challenges.

Some methods have been applied to causal problems that are extremely relevant for robotics. Volodin et al. [317] is a method for developing a causal model for reinforcement learning and fixing errors in the causal model. They perform these fixes using interventions and test them on a grid-world environment. This is well-suited for extension to a robot domain because a robot interacts with the world and can perform interventions.

The insights from [236], which seeks to identify which decisions are the most important/have the most impact on success of a policy (discussed in Sect. 4.2), could be used to identify key decision points. This could be applied before, during, or after the creation of a causal model and could be helpful in the difficult task of building a useful causal model from observations of a large state space over many time steps.

Reward Decomposition:

In the case of a human-robot task that has a reward function that can be decomposed into its components, the RARE framework [292] estimates the rewards missing from a user's comprehension of the domain's true reward function. The existing literature does not consider cases where the user has imagined a reward component that is not present in the true reward function at all. The method covers the user missing a true reward, but not the case where the user erroneously includes an incorrect or nonexistent reward signal in their comprehension of the domain. This presents a valuable future direction.

Additionally, any of the reward decomposition methods, including the standard ones, would be well served by incorporating methods for identifying reward functions or even taking a given complex or scalar reward function and decomposing it into meaningful components. This would help remove an upfront human cost in creating these explanations.

Visualizations:

Many existing categories of explanation methods in this survey lend themselves to the development of useful visualizations, including Dimension Reduction (Sect. 3.4.1), Causal Methods (Sect. 3.12), and methods involving graphs (Sect. 3.16) or hierarchies (Sect. 3.9).

Instruction Following:

The approaches to learning by instructions in [268, 310] have clear relevance to safety, although neither paper makes this connection. In particular, [310] represents actions as code functions, and the robot chooses which function to execute as an action. This has been demonstrated in a real-world robot. The safety application comes when we understand that each individual function can be unit tested and their compositions can be tested, both theoretically and empirically. Thus, a policy composed of such functions can be analyzed and verified in ways that a raw neural network policy could never be. Exploring this idea would be a valuable future direction. Additionally, representing this policy as a composition of code could allow for customization and modification, another direction to pursue.

We see several approaches for learning to create explanations for actions based on human instructions to perform those actions [57, 93, 94]. (See additional papers on instruction following without explicit explanations as well: [142, 145, 268, 310].) These approaches could be combined with work on natural language processing analysis of visual images to more efficiently and accurately propose explanations for a robot's chosen actions that are not close to any particular instruction a human has previously given.

A user study in [156] explores the effects of temporarily modifying a reinforcement learning-based policy in real time. When instructions are followed only sometimes, or after a delay, humans get frustrated. In addition to existing techniques that might seek to learn from such errors [57] when they are known, it would be worthwhile to work on techniques to mitigate these errors in the first place. At the very least, attempting to detect failure to follow instructions and offering an additional explanation to the human is another opportunity for useful explanation.

There is work on using human feedback to correct a robot's current policy [183], discussed in Sect. 3.15. This use of knowledge limits is an interesting contrast to the case of an agent using counterfactuals to justify why its plan is superior to a human's suggestion [286] in Sect. 3.3.2, where human suggestions were rejected based on what the algorithm knew. A productive future direction would be to combine these two ideas to create an agent that receives human feedback and, based on its level of uncertainty either incorporates it or rejects it with an explanation while also giving general explanations of its behavior based on human language. This would allow for more complex human-robot interaction while moving closer to a real-world scenario where a human might have expertise in some areas but not others. A robot that could intelligently figure out when to overrule a human or not would be useful. (Although, naturally, the idea might scare some people, and some would want to allow the human to override the override.)

The delivery robot simulation in [183], with the focus on human feedback, is ripe for combining with research into legibility/readability (Sects. 3.17 and 2.3). Since the goal is for the human to observe and give feedback, a robot that is better able to communicate intentions and goals to a human could receive this feedback quicker, facilitating the entire process. Simply increasing the complexity is also worthwhile. The simulation is a simple task; the robot is moving, and the human sees it move. More complex applications could involve more serious concerns.

We also bring up [217], which is similar to instruction following reinforcement learning, but the reward signal is replaced with a description. This could be built upon.

Symbolic Methods:

Trees containing symbolic functions are used in [269] to map from raw states to a scalar reward. In the case of the sparse reward environment, this intermediate reward can aid in learning. This paper develops the trees using evolutionary methods. Other means of creating trees could also be explored, or even other methods of using logic to attain intermediate rewards. Additionally, [269] tests this approach in simulation, and it is worth applying it on a real-world robot.

New techniques for symbolic regression have come out recently from [209, 314], which could be applied to robotics.

See also the comments on extending [180] in the Decision Tree subsection above.

Legibility and Readability:

Legibility and readability (Sects. 2.3 and 3.17) are important concepts in robotics. Given that most of the reinforcement learning field is focused on non-robotics work, it is no surprise that these ideas remain understudied in the context of reinforcement learning (and even machine learning in general). Any project that seeks to introduce these concepts to an existing reinforcement learning method or to combine them with any other explainable method presented in this survey would be worthwhile.

To take it even farther, one could build on [158], which does not involve reinforcement learning but instead involves signaling information to allies while hiding information from enemies. This method requires full knowledge of sensor models and prior knowledge of allies and enemies. It would be interesting to extend this idea to a robotics domain where reinforcement learning is applied, potentially with more uncertainty in the environment.

Additional Directions:

Most of the methods described in this work are concerned with explaining a single agent. As interest grows in multi-agent reinforcement learning (MARL), work is being done on explainable MARL [39]. Extending these types of ideas for multi-agent robotics would be a logical next step.

Robots will often be used by non-experts. Part of explainability is not just understanding what the robot understands but correcting it. Waveren et al. [308] introduces a method for human laypersons to correct the errors in an RL agent. It would be worthwhile to extend these ideas into robotics domains.

5.2 Conclusion

In this broad survey, we inventory the state of the art in Explainable and Interpretable Reinforcement Learning for Robotics. We describe 12 attributes that can be used to compare explainability techniques, including a discussion of how terminology is used in the field. We elaborate on 42 categories and subcategories of XRL techniques relevant to robotics, discussing the state of the art in each and identifying where they fall on the attributes above. We provide a discussion of these topics and offer a multitude of suggestions for future directions for research.

The intersection of topics in this survey touches on subjects that are becoming extremely relevant in the laboratory and the real world. As artificial intelligence in robots becomes more powerful, we hope this helps interested researchers to continue to make advances in helping humans to understand, use, interact with, and collaborate with these systems.

References

1. Moloud Abdar, Farhad Pourpanah, Sadiq Hussain, Dana Rezazadegan, Li Liu, Mohammad Ghavamzadeh, Paul Fieguth, Xiaochun Cao, Abbas Khosravi, U Rajendra Acharya, et al. A review of uncertainty quantification in deep learning: Techniques, applications and challenges. *Information Fusion*, 2021.
2. Aastha Acharya, Rebecca Russell, and Nisar R Ahmed. Explaining conditions for reinforcement learning behaviors from real and imagined data. *Workshop on Challenges of Real-World RL at the 34th Conference on Neural Information Processing Systems (NeurIPS 2020)*, 2020.
3. Joshua Achiam, David Held, Aviv Tamar, and Pieter Abbeel. Constrained policy optimization. In *International Conference on Machine Learning*, pages 22–31. PMLR, 2017.
4. Amina Adadi and Mohammed Berrada. Peeking inside the black-box: a survey on explainable artificial intelligence (xai). *IEEE access*, 6:52138–52160, 2018.
5. Alekh Agarwal, Sarah Bird, Markus Cozowicz, Luong Hoang, John Langford, Stephen Lee, Jiaji Li, Dan Melamed, Gal Oshri, Oswaldo Ribas, et al. Making contextual decisions with low technical debt. *arXiv preprint* arXiv:1606.03966, 2016.
6. Ossama Ahmed, Frederik Träuble, Anirudh Goyal, Alexander Neitz, Manuel Wüthrich, Yoshua Bengio, Bernhard Schölkopf, and Stefan Bauer. Causalworld: A robotic manipulation benchmark for causal structure and transfer learning, 2020.
7. Lun Ai, Stephen H Muggleton, Céline Hocquette, Mark Gromowski, and Ute Schmid. Beneficial and harmful explanatory machine learning. *Machine Learning*, 110(4):695–721, 2021.
8. Riad Akrour, Davide Tateo, and Jan Peters. Towards reinforcement learning of human readable policies, 2019.
9. Riad Akrour, Davide Tateo, and Jan Peters. Reinforcement learning from a mixture of interpretable experts. *arXiv preprint* arXiv:2006.05911, 2020.
10. Alnour Alharin, Thanh-Nam Doan, and Mina Sartipi. Reinforcement learning interpretation methods: A survey. *IEEE Access*, 8:171058–171077, 2020.
11. Dan Amir and Ofra Amir. Highlights: Summarizing agent behavior to people. In *Proceedings of the 17th International Conference on Autonomous Agents and MultiAgent Systems*, pages 1168–1176, 2018.

12. Ofra Amir, Finale Doshi-Velez, and David Sarne. Agent strategy summarization. In *Proceedings of the 17th International Conference on Autonomous Agents and MultiAgent Systems*, pages 1203–1207, 2018.
13. Ofra Amir, Finale Doshi-Velez, and David Sarne. Summarizing agent strategies. *Autonomous Agents and Multi-Agent Systems*, 33(5):628–644, 2019.
14. Yotam Amitai and Ofra Amir. "i don't think so": Summarizing policy disagreements for agent comparison. *Proceedings of the AAAI Conference on Artificial Intelligence*, 36(5):5269–5276, Jun. 2022.
15. Andrew Anderson, Jonathan Dodge, Amrita Sadarangani, Zoe Juozapaitis, Evan Newman, Jed Irvine, Souti Chattopadhyay, Matthew Olson, Alan Fern, and Margaret Burnett. Mental models of mere mortals with explanations of reinforcement learning. *ACM Transactions on Interactive Intelligent Systems (TiiS)*, 10(2):1–37, 2020.
16. Greg Anderson, Abhinav Verma, Isil Dillig, and Swarat Chaudhuri. Neurosymbolic reinforcement learning with formally verified exploration. *Advances in neural information processing systems*, 33:6172–6183, 2020.
17. Jonas Andrulis, Ole Meyer, Grégory Schott, Samuel Weinbach, and Volker Gruhn. Domain-level explainability–a challenge for creating trust in superhuman ai strategies. *arXiv preprint* arXiv:2011.06665, 2020.
18. Marcin Andrychowicz, Filip Wolski, Alex Ray, Jonas Schneider, Rachel Fong, Peter Welinder, Bob McGrew, Josh Tobin, OpenAI Pieter Abbeel, and Wojciech Zaremba. Hindsight experience replay. *Advances in neural information processing systems*, 30, 2017.
19. OpenAI: Marcin Andrychowicz, Bowen Baker, Maciek Chociej, Rafal Jozefowicz, Bob McGrew, Jakub Pachocki, Arthur Petron, Matthias Plappert, Glenn Powell, Alex Ray, et al. Learning dexterous in-hand manipulation. *The International Journal of Robotics Research*, 39(1):3–20, 2020.
20. Sule Anjomshoae, Amro Najjar, Davide Calvaresi, and Kary Främling. Explainable agents and robots: Results from a systematic literature review. In *18th International Conference on Autonomous Agents and Multiagent Systems (AAMAS 2019), Montreal, Canada, May 13–17, 2019*, pages 1078–1088. International Foundation for Autonomous Agents and Multiagent Systems, 2019.
21. Raghuram Mandyam Annasamy and Katia Sycara. Towards better interpretability in deep q-networks. In *Proceedings of the AAAI Conference on Artificial Intelligence*, volume 33, pages 4561–4569, 2019.
22. Saurabh Arora and Prashant Doshi. A survey of inverse reinforcement learning: Challenges, methods and progress. *Artificial Intelligence*, page 103500, 2021.
23. Alejandro Barredo Arrieta, Natalia Díaz-Rodríguez, Javier Del Ser, Adrien Bennetot, Siham Tabik, Alberto Barbado, Salvador García, Sergio Gil-López, Daniel Molina, Richard Benjamins, et al. Explainable artificial intelligence (xai): Concepts, taxonomies, opportunities and challenges toward responsible ai. *Information Fusion*, 58:82–115, 2020.
24. Akanksha Atrey, Kaleigh Clary, and David Jensen. Exploratory not explanatory: Counterfactual analysis of saliency maps for deep reinforcement learning. *ICLR 2020*, 2020.
25. Akhil Bagaria, Jason Crowley, Jing Wei Nicholas Lim, and George Konidaris. Skill discovery for exploration and planning using deep skill graphs, 2020.
26. Dzmitry Bahdanau, Kyunghyun Cho, and Yoshua Bengio. Neural machine translation by jointly learning to align and translate. *ICLR 2015*, 2015.
27. Arindam Banerjee and Sugato Basu. Topic models over text streams: A study of batch and online unsupervised learning. In *Proceedings of the 2007 SIAM International Conference on Data Mining*, pages 431–436. SIAM, 2007.
28. Horace B Barlow. Unsupervised learning. *Neural computation*, 1(3):295–311, 1989.

29. Yavar Bathaee. The artificial intelligence black box and the failure of intent and causation. *Harv. JL & Tech.*, 31:889, 2017.
30. Seth Baum. A survey of artificial general intelligence projects for ethics, risk, and policy. *Global Catastrophic Risk Institute Working Paper*, pages 17–1, 2017.
31. Yoshua Bengio, Aaron Courville, and Pascal Vincent. Representation learning: A review and new perspectives. *IEEE transactions on pattern analysis and machine intelligence*, 35(8):1798–1828, 2013.
32. Tarek R Besold, Artur d'Avila Garcez, Sebastian Bader, Howard Bowman, Pedro Domingos, Pascal Hitzler, Kai-Uwe Kühnberger, Luis C Lamb, Daniel Lowd, Priscila Machado Vieira Lima, et al. Neural-symbolic learning and reasoning: A survey and interpretation. *arXiv preprint* arXiv:1711.03902, 2017.
33. JG Bever, WL Kilmer, WS Mc Culloch, R Moreno-Diaz, and LL Sutro. Development of visual, contact and decision subsystems for a mars rover, july 1966-january 1967. Technical report, Massachusetts Institute of Technology, Instrumentation Laboratory, 1967.
34. Benjamin Beyret, Ali Shafti, and A Aldo Faisal. Dot-to-dot: Explainable hierarchical reinforcement learning for robotic manipulation. In *2019 IEEE/RSJ International Conference on Intelligent Robots and Systems (IROS)*, pages 5014–5019. IEEE, 2019.
35. Umang Bhatt, Alice Xiang, Shubham Sharma, Adrian Weller, Ankur Taly, Yunhan Jia, Joydeep Ghosh, Ruchir Puri, José MF Moura, and Peter Eckersley. Explainable machine learning in deployment. In *Proceedings of the 2020 Conference on Fairness, Accountability, and Transparency*, pages 648–657, 2020.
36. Manuel Bied and Mohamed Chetouani. Integrating an observer in interactive reinforcement learning to learn legible trajectories. In *2020 29th IEEE International Conference on Robot and Human Interactive Communication (RO-MAN)*, pages 760–767. IEEE, 2020.
37. Bruno Blanchet, Patrick Cousot, Radhia Cousot, Jérome Feret, Laurent Mauborgne, Antoine Miné, David Monniaux, and Xavier Rival. A static analyzer for large safety-critical software. In *Proceedings of the ACM SIGPLAN 2003 conference on Programming language design and implementation*, pages 196–207, 2003.
38. Kayla Boggess, Shenghui Chen, and Lu Feng. Towards personalized explanation of robot path planning via user feedback. *arXiv preprint* arXiv:2011.00524, 2020.
39. Kayla Boggess, Sarit Kraus, and Lu Feng. Toward policy explanations for multi-agent reinforcement learning. *arXiv preprint* arXiv:2204.12568, 2022.
40. Nicolò Botteghi, Khaled Alaa, Mannes Poel, Beril Sirmacek, Christoph Brune, Abeje Mersha, and Stefano Stramigioli. Low dimensional state representation learning with robotics priors in continuous action spaces. *arXiv preprint* arXiv:2107.01667, 2021.
41. Nicolò Botteghi, Ruben Obbink, Daan Geijs, Mannes Poel, Beril Sirmacek, Christoph Brune, Abeje Mersha, and Stefano Stramigioli. Low dimensional state representation learning with reward-shaped priors. In *2020 25th International Conference on Pattern Recognition (ICPR)*, pages 3736–3743. IEEE, 2021.
42. Greg Brockman, Vicki Cheung, Ludwig Pettersson, Jonas Schneider, John Schulman, Jie Tang, and Wojciech Zaremba. Openai gym. *arXiv preprint* arXiv:1606.01540, 2016.
43. David A Broniatowski et al. Psychological foundations of explainability and interpretability in artificial intelligence. *NIST: National Institute of Standards and Technology, US Department of Commerce*, 2021.
44. Alexander Brown and Marek Petrik. Interpretable reinforcement learning with ensemble methods. *arXiv preprint* arXiv:1809.06995, 2018.
45. Lindsey Jacquelyn Byom and Bilge Mutlu. Theory of mind: Mechanisms, methods, and new directions. *Frontiers in human neuroscience*, 7:413, 2013.
46. Roberta Calegari, Giovanni Ciatto, and Andrea Omicini. On the integration of symbolic and sub-symbolic techniques for xai: A survey. *Intelligenza Artificiale*, 14(1):7–32, 2020.

47. Diogo V Carvalho, Eduardo M Pereira, and Jaime S Cardoso. Machine learning interpretability: A survey on methods and metrics. *Electronics*, 8(8):832, 2019.
48. Tathagata Chakraborti, Anagha Kulkarni, Sarath Sreedharan, David E Smith, and Subbarao Kambhampati. Explicability? legibility? predictability? transparency? privacy? security? the emerging landscape of interpretable agent behavior. In *Proceedings of the international conference on automated planning and scheduling*, volume 29, pages 86–96, 2019.
49. Tathagata Chakraborti, Sarath Sreedharan, and Subbarao Kambhampati. Explicability versus explanations in human-aware planning. In *Proceedings of the 17th International Conference on Autonomous Agents and MultiAgent Systems*, pages 2180–2182, 2018.
50. Tathagata Chakraborti, Sarath Sreedharan, Yu Zhang, and Subbarao Kambhampati. Plan explanations as model reconciliation: Moving beyond explanation as soliloquy. *arXiv preprint* arXiv:1701.08317, 2017.
51. Supriyo Chakraborty, Richard Tomsett, Ramya Raghavendra, Daniel Harborne, Moustafa Alzantot, Federico Cerutti, Mani Srivastava, Alun Preece, Simon Julier, Raghuveer M Rao, et al. Interpretability of deep learning models: A survey of results. In *2017 IEEE smartworld, ubiquitous intelligence & computing, advanced & trusted computed, scalable computing & communications, cloud & big data computing, Internet of people and smart city innovation (smartworld/SCALCOM/UIC/ATC/CBDcom/IOP/SCI)*, pages 1–6. IEEE, 2017.
52. Constantinos Chamzas, Martina Lippi, Michael C Welle, Anastasiia Varava, Alessandro Marino, Lydia E Kavraki, and Danica Kragic. State representations in robotics: Identifying relevant factors of variation using weak supervision. In *Robot Learning Workshop, Neurips*, 2020.
53. Mai Lee Chang, Reymundo A Gutierrez, Priyanka Khante, Elaine Schaertl Short, and Andrea Lockerd Thomaz. Effects of integrated intent recognition and communication on human-robot collaboration. In *2018 IEEE/RSJ International Conference on Intelligent Robots and Systems (IROS)*, pages 3381–3386. IEEE, 2018.
54. Shenghui Chen, Kayla Boggess, and Lu Feng. Towards transparent robotic planning via contrastive explanations. In *2020 IEEE/RSJ International Conference on Intelligent Robots and Systems (IROS)*, pages 6593–6598. IEEE, 2020.
55. Hui Chia. In machines we trust: Are robo-advisers more trustworthy than human financial advisers? *Law, Tech. & Hum.*, 1:129, 2019.
56. Brian Christian. *The Alignment Problem: Machine Learning and Human Values*. WW Norton & Company, 2020.
57. Geoffrey Cideron, Mathieu Seurin, Florian Strub, and Olivier Pietquin. Higher: Improving instruction following with hindsight generation for experience replay. In *2020 IEEE Symposium Series on Computational Intelligence (SSCI)*, pages 225–232. IEEE, 2020.
58. Youri Coppens, Kyriakos Efthymiadis, Tom Lenaerts, Ann Nowé, Tim Miller, Rosina Weber, and Daniele Magazzeni. Distilling deep reinforcement learning policies in soft decision trees. In *Proceedings of the IJCAI 2019 Workshop on Explainable Artificial Intelligence*, pages 1–6, 2019.
59. Davide Corsi, Raz Yerushalmi, Guy Amir, Alessandro Farinelli, David Harel, and Guy Katz. Constrained reinforcement learning for robotics via scenario-based programming. *arXiv preprint* arXiv:2206.09603, 2022.
60. Andrew Critch and David Krueger. Ai research considerations for human existential safety (arches). *arXiv preprint* arXiv:2006.04948, 2020.
61. Francisco Cruz, Richard Dazeley, and Peter Vamplew. Memory-based explainable reinforcement learning. In *Australasian Joint Conference on Artificial Intelligence*, pages 66–77. Springer, 2019.
62. Francisco Cruz, Richard Dazeley, Peter Vamplew, and Ithan Moreira. Explainable robotic systems: Understanding goal-driven actions in a reinforcement learning scenario. *Neural Computing and Applications*, pages 1–18, 2021.

References

63. Francisco Cruz, Charlotte Young, Richard Dazeley, and Peter Vamplew. Evaluating human-like explanations for robot actions in reinforcement learning scenarios. *IEEE/RSJ International Conference on Intelligent Robots and Systems (IROS)*, 2022.
64. Pádraig Cunningham, Matthieu Cord, and Sarah Jane Delany. Supervised learning. In *Machine learning techniques for multimedia*, pages 21–49. Springer, 2008.
65. Leonardo Lucio Custode and Giovanni Iacca. Evolutionary learning of interpretable decision trees. *arXiv preprint* arXiv:2012.07723, 2020.
66. Tianhong Dai, Kai Arulkumaran, Tamara Gerbert, Samyakh Tukra, Feryal Behbahani, and Anil Anthony Bharath. Analysing deep reinforcement learning agents trained with domain randomisation. *Neurocomputing*, 493:143–165, 2022.
67. Wang-Zhou Dai and Stephen H Muggleton. Abductive knowledge induction from raw data. *arXiv preprint* arXiv:2010.03514, 2020.
68. Devleena Das, Siddhartha Banerjee, and Sonia Chernova. Explainable ai for robot failures: Generating explanations that improve user assistance in fault recovery. In *Proceedings of the 2021 ACM/IEEE International Conference on Human-Robot Interaction*, pages 351–360, 2021.
69. Omid Davoodi and Majid Komeili. Feature-based interpretable reinforcement learning based on state-transition models. In *2021 IEEE International Conference on Systems, Man, and Cybernetics (SMC)*, pages 301–308. IEEE, 2021.
70. Michiel de Jong, Kevin Zhang, Aaron M Roth, Travers Rhodes, Robin Schmucker, Chenghui Zhou, Sofia Ferreira, João Cartucho, and Manuela Veloso. Towards a robust interactive and learning social robot. In *Proceedings of the 17th International Conference on Autonomous Agents and MultiAgent Systems*, pages 883–891, 2018.
71. Leonardo De Moura and Nikolaj Bjørner. Z3: An efficient smt solver. In *International conference on Tools and Algorithms for the Construction and Analysis of Systems*, pages 337–340. Springer, 2008.
72. Thomas Dean, James Allen, and Yiannis Aloimonos. *Artificial intelligence: theory and practice*. Benjamin-Cummings Publishing Co., Inc., 1995.
73. Shuby Deshpande, Benjamin Eysenbach, and Jeff Schneider. Interactive visualization for debugging rl. *arXiv preprint* arXiv:2008.07331, 2020.
74. Yashesh Dhebar and Kalyanmoy Deb. Interpretable rule discovery through bilevel optimization of split-rules of nonlinear decision trees for classification problems. *IEEE Transactions on Cybernetics*, 2020.
75. Yashesh Dhebar, Kalyanmoy Deb, Subramanya Nageshrao, Ling Zhu, and Dimitar Filev. Toward interpretable-ai policies using evolutionary nonlinear decision trees for discrete-action systems. *IEEE Transactions on Cybernetics*, 2022.
76. Derek Doran, Sarah Schulz, and Tarek R Besold. What does explainable ai really mean? a new conceptualization of perspectives. *arXiv preprint* arXiv:1710.00794, 2017.
77. Finale Doshi-Velez and Been Kim. Towards a rigorous science of interpretable machine learning. *arXiv preprint* arXiv:1702.08608, 2017.
78. Filip Karlo Došilović, Mario Brčić, and Nikica Hlupić. Explainable artificial intelligence: A survey. In *2018 41st International convention on information and communication technology, electronics and microelectronics (MIPRO)*, pages 0210–0215. IEEE, 2018.
79. Alexey Dosovitskiy, German Ros, Felipe Codevilla, Antonio Lopez, and Vladlen Koltun. Carla: An open urban driving simulator. In *Conference on robot learning*, pages 1–16. PMLR, 2017.
80. Anca D Dragan, Kenton CT Lee, and Siddhartha S Srinivasa. Legibility and predictability of robot motion. In *2013 8th ACM/IEEE International Conference on Human-Robot Interaction (HRI)*, pages 301–308. IEEE, 2013.
81. Gabriel Dulac-Arnold, Daniel Mankowitz, and Todd Hester. Challenges of real-world reinforcement learning. *arXiv preprint* arXiv:1904.12901, 2019.

82. Francesco Esposito, Christian Pek, Michael C Welle, and Danica Kragic. Learning task constraints in visual-action planning from demonstrations. In *IEEE Int. Conf. on Robot and Human Interactive Communication*, 2021.
83. Richard Evans, Matko Bošnjak, Lars Buesing, Kevin Ellis, David Pfau, Pushmeet Kohli, and Marek Sergot. Making sense of raw input. *Artificial Intelligence*, 299:103521, 2021.
84. Benjamin Eysenbach, Abhishek Gupta, Julian Ibarz, and Sergey Levine. Diversity is all you need: Learning skills without a reward function. *arXiv preprint* arXiv:1802.06070, 2018.
85. Kanvaly Fadiga, Etienne Houzé, Ada Diaconescu, and Jean-Louis Dessalles. To do or not to do: Finding causal relations in smart homes. In *2021 IEEE International Conference on Autonomic Computing and Self-Organizing Systems (ACSOS)*, pages 110–119. IEEE, 2021.
86. Bin Fang, Shidong Jia, Di Guo, Muhua Xu, Shuhuan Wen, and Fuchun Sun. Survey of imitation learning for robotic manipulation. *International Journal of Intelligent Robotics and Applications*, 3(4):362–369, 2019.
87. Juliana J Ferreira and Mateus S Monteiro. What are people doing about xai user experience? a survey on ai explainability research and practice. In *International Conference on Human-Computer Interaction*, pages 56–73. Springer, 2020.
88. Jaime F Fisac, Chang Liu, Jessica B Hamrick, Shankar Sastry, J Karl Hedrick, Thomas L Griffiths, and Anca D Dragan. Generating plans that predict themselves. In *Algorithmic Foundations of Robotics XII*, pages 144–159. Springer, 2020.
89. Ruth C Fong and Andrea Vedaldi. Interpretable explanations of black boxes by meaningful perturbation. In *Proceedings of the IEEE International Conference on Computer Vision*, pages 3429–3437, 2017.
90. Abraham Frandsen, Rong Ge, and Holden Lee. Extracting latent state representations with linear dynamics from rich observations. In *International Conference on Machine Learning*, pages 6705–6725. PMLR, 2022.
91. Alex A Freitas. Comprehensible classification models: a position paper. *ACM SIGKDD explorations newsletter*, 15(1):1–10, 2014.
92. Yasuhiro Fujita, Kota Uenishi, Avinash Ummadisingu, Prabhat Nagarajan, Shimpei Masuda, and Mario Ynocente Castro. Distributed reinforcement learning of targeted grasping with active vision for mobile manipulators. In *2020 IEEE/RSJ International Conference on Intelligent Robots and Systems (IROS)*, pages 9712–9719. IEEE, 2020.
93. Yosuke Fukuchi, Masahiko Osawa, Hiroshi Yamakawa, and Michita Imai. Application of instruction-based behavior explanation to a reinforcement learning agent with changing policy. In *International Conference on Neural Information Processing*, pages 100–108. Springer, 2017.
94. Yosuke Fukuchi, Masahiko Osawa, Hiroshi Yamakawa, and Michita Imai. Autonomous self-explanation of behavior for interactive reinforcement learning agents. In *Proceedings of the 5th International Conference on Human Agent Interaction*, pages 97–101, 2017.
95. Briti Gangopadhyay and Pallab Dasgupta. Counterexample guided rl policy refinement using bayesian optimization. *Advances in Neural Information Processing Systems*, 34:22783–22794, 2021.
96. Briti Gangopadhyay, Harshit Soora, and Pallab Dasgupta. Hierarchical program-triggered reinforcement learning agents for automated driving. *IEEE Transactions on Intelligent Transportation Systems*, 2021.
97. Xiaofeng Gao, Ran Gong, Tianmin Shu, Xu Xie, Shu Wang, and Song-Chun Zhu. Vrkitchen: an interactive 3d virtual environment for task-oriented learning. *arXiv preprint* arXiv:1903.05757, 2019.
98. Artur d'Avila Garcez, Sebastian Bader, Howard Bowman, Luis C Lamb, Leo de Penning, BV Illuminoo, Hoifung Poon, and COPPE Gerson Zaverucha. Neural-symbolic learning and reasoning: A survey and interpretation. *Neuro-Symbolic Artificial Intelligence: The State of the Art*, 342:1, 2022.

99. Marta Garnelo and Murray Shanahan. Reconciling deep learning with symbolic artificial intelligence: representing objects and relations. *Current Opinion in Behavioral Sciences*, 29:17–23, 2019.
100. Abhiroop Ghosh, Yashesh Dhebar, Ritam Guha, Kalyanmoy Deb, Subramanya Nageshrao, Ling Zhu, Eric Tseng, and Dimitar Filev. Interpretable ai agent through nonlinear decision trees for lane change problem. In *2021 IEEE Symposium Series on Computational Intelligence (SSCI)*, pages 01–08. IEEE, 2021.
101. Leilani H Gilpin, David Bau, Ben Z Yuan, Ayesha Bajwa, Michael Specter, and Lalana Kagal. Explaining explanations: An overview of interpretability of machine learning. In *2018 IEEE 5th International Conference on data science and advanced analytics (DSAA)*, pages 80–89. IEEE, 2018.
102. Claire Glanois, Paul Weng, Matthieu Zimmer, Dong Li, Tianpei Yang, Jianye Hao, and Wulong Liu. A survey on interpretable reinforcement learning. *arXiv preprint* arXiv:2112.13112, 2021.
103. Alyssa Glass, Deborah L McGuinness, and Michael Wolverton. Toward establishing trust in adaptive agents. In *Proceedings of the 13th international conference on Intelligent user interfaces*, pages 227–236, 2008.
104. Samuel Greydanus, Anurag Koul, Jonathan Dodge, and Alan Fern. Visualizing and understanding atari agents. In *International Conference on Machine Learning*, pages 1792–1801. PMLR, 2018.
105. Ross Gruetzemacher, David Paradice, and Kang Bok Lee. Forecasting transformative ai: An expert survey. *arXiv preprint* arXiv:1901.08579, 2019.
106. Riccardo Guidotti, Anna Monreale, Salvatore Ruggieri, Franco Turini, Fosca Giannotti, and Dino Pedreschi. A survey of methods for explaining black box models. *ACM computing surveys (CSUR)*, 51(5):1–42, 2018.
107. David Gunning, Mark Stefik, Jaesik Choi, Timothy Miller, Simone Stumpf, and Guang-Zhong Yang. Xai-explainable artificial intelligence. *Science Robotics*, 4(37), 2019.
108. David Gunning, Eric Vorm, Jennifer Yunyan Wang, and Matt Turek. Darpa's explainable ai (xai) program: A retrospective, 2021.
109. Suna Sihang Guo, Ruohan Zhang, Bo Liu, Yifeng Zhu, Dana Ballard, Mary Hayhoe, and Peter Stone. Machine versus human attention in deep reinforcement learning tasks. *Advances in Neural Information Processing Systems*, 34:25370–25385, 2021.
110. Abhishek Gupta, Ahmed Shaharyar Khwaja, Alagan Anpalagan, Ling Guan, and Bala Venkatesh. Policy-gradient and actor-critic based state representation learning for safe driving of autonomous vehicles. *Sensors*, 20(21):5991, 2020.
111. Piyush Gupta, Nikaash Puri, Sukriti Verma, Sameer Singh, Dhruv Kayastha, Shripad Deshmukh, and Balaji Krishnamurthy. Explain your move: Understanding agent actions using focused feature saliency. *arXiv preprint* arXiv:1912.12191, 2019.
112. Lawrence O Hall, Nitesh Chawla, and Kevin W Bowyer. Decision tree learning on very large data sets. In *SMC'98 Conference Proceedings. 1998 IEEE International Conference on Systems, Man, and Cybernetics (Cat. No. 98CH36218)*, volume 3, pages 2579–2584. IEEE, 1998.
113. Patrick Hall, Navdeep Gill, and Nicholas Schmidt. Proposed guidelines for the responsible use of explainable machine learning. *Robust AI in Financial Services Workshop at the 33rd Conference on Neural Information Processing Systems*, 2019.
114. Ronan Hamon, Henrik Junklewitz, and Ignacio Sanchez. Robustness and explainability of artificial intelligence. *Publications Office of the European Union*, 2020.
115. Zhao Han, Phillips Elizabeth, and Hollay A Yanco. The need for verbal robot explanations and how people would like a robot to explain itself. *ACM Trans. Hum.-Robot Interact*, 1(1), 2021.
116. Bradley Hayes and Julie A Shah. Improving robot controller transparency through autonomous policy explanation. In *2017 12th ACM/IEEE International Conference on Human-Robot Interaction (HRI*, pages 303–312. IEEE, 2017.

117. Lei He, Nabil Aouf, and Bifeng Song. Explainable deep reinforcement learning for uav autonomous path planning. *Aerospace science and technology*, 118:107052, 2021.
118. Lei He, Nabil Aouf, and Bifeng Song. Explainable deep reinforcement learning for uav autonomous path planning. *Aerospace science and technology*, 118:107052, 2021.
119. Daniel Hein, Steffen Udluft, and Thomas A Runkler. Interpretable policies for reinforcement learning by genetic programming. *Engineering Applications of Artificial Intelligence*, 76:158–169, 2018.
120. Donald Hejna, Lerrel Pinto, and Pieter Abbeel. Hierarchically decoupled imitation for morphological transfer. In *International Conference on Machine Learning*, pages 4159–4171. PMLR, 2020.
121. Malte Helmert and Héctor Geffner. Unifying the causal graph and additive heuristics. In *ICAPS*, pages 140–147, 2008.
122. Adam J Hepworth, Daniel P Baxter, Aya Hussein, Kate J Yaxley, Essam Debie, and Hussein A Abbass. Human-swarm-teaming transparency and trust architecture. *IEEE/CAA Journal of Automatica Sinica*, 2020.
123. José Hernández-Orallo, Fernando Martínez-Plumed, Shahar Avin, et al. Surveying safety-relevant ai characteristics. In *SafeAI@ AAAI*, 2019.
124. Alexandre Heuillet, Fabien Couthouis, and Natalia Díaz-Rodríguez. Explainability in deep reinforcement learning. *Knowledge-Based Systems*, 214:106685, 2021.
125. Thomas Hickling, Nabil Aouf, and Phillippa Spencer. Robust adversarial attacks detection based on explainable deep reinforcement learning for uav guidance and planning. *arXiv preprint arXiv:2206.02670*, 2022.
126. Thomas Hofmann. The cluster-abstraction model: Unsupervised learning of topic hierarchies from text data. In *IJCAI*, volume 99, pages 682–687. Citeseer, 1999.
127. Juraj Holas and Igor Farkaš. Adaptive skill acquisition in hierarchical reinforcement learning. In *International Conference on Artificial Neural Networks*, pages 383–394. Springer, 2020.
128. Jarrett Holtz, Simon Andrews, Arjun Guha, and Joydeep Biswas. Iterative program synthesis for adaptable social navigation. In *2021 IEEE/RSJ International Conference on Intelligent Robots and Systems (IROS)*, pages 6256–6261. IEEE, 2021.
129. Sandy H Huang, Kush Bhatia, Pieter Abbeel, and Anca D Dragan. Leveraging critical states to develop trust. In *RSS 2017 Workshop: Morality and Social Trust in Autonomous Robots*, 2017.
130. Sandy H Huang, Kush Bhatia, Pieter Abbeel, and Anca D Dragan. Establishing appropriate trust via critical states. In *2018 IEEE/RSJ International Conference on Intelligent Robots and Systems (IROS)*, pages 3929–3936. IEEE, 2018.
131. Sandy H Huang, David Held, Pieter Abbeel, and Anca D Dragan. Enabling robots to communicate their objectives. *Autonomous Robots*, 43(2):309–326, 2019.
132. Tobias Huber, Benedikt Limmer, and Elisabeth André. Benchmarking perturbation-based saliency maps for explaining deep reinforcement learning agents. *arXiv preprint arXiv:2101.07312*, 2021.
133. Tobias Huber, Katharina Weitz, Elisabeth André, and Ofra Amir. Local and global explanations of agent behavior: Integrating strategy summaries with saliency maps. *Artificial Intelligence*, 301:103571, 2021.
134. Ahmed Hussein, Mohamed Medhat Gaber, Eyad Elyan, and Chrisina Jayne. Imitation learning: A survey of learning methods. *ACM Computing Surveys (CSUR)*, 50(2):1–35, 2017.
135. Julian Ibarz, Jie Tan, Chelsea Finn, Mrinal Kalakrishnan, Peter Pastor, and Sergey Levine. How to train your robot with deep reinforcement learning: lessons we have learned. *The International Journal of Robotics Research*, 40(4-5):698–721, 2021.
136. Brett W Israelsen and Nisar R Ahmed. "dave... i can assure you... that it's going to be all right..." a definition, case for, and survey of algorithmic assurances in human-autonomy trust relationships. *ACM Computing Surveys (CSUR)*, 51(6):1–37, 2019.

137. Alessandro Iucci. Explainable reinforcement learning for risk mitigation in human-robot collaboration scenarios, 2021.
138. Maksims Ivanovs, Roberts Kadikis, and Kaspars Ozols. Perturbation-based methods for explaining deep neural networks: A survey. *Pattern Recognition Letters*, 150:228–234, 2021.
139. Utkarshani Jaimini and Amit Sheth. Causalkg: Causal knowledge graph explainability using interventional and counterfactual reasoning. *IEEE Internet Computing*, 26(1):43–50, 2022.
140. Miguel Jaques, Michael Burke, and Timothy M Hospedales. Newtonianvae: Proportional control and goal identification from pixels via physical latent spaces. In *Proceedings of the IEEE/CVF Conference on Computer Vision and Pattern Recognition*, pages 4454–4463, 2021.
141. Xia Jiang, Jian Zhang, and Bo Wang. Energy-efficient driving for adaptive traffic signal control environment via explainable reinforcement learning. *Applied Sciences*, 12(11):5380, 2022.
142. Yiding Jiang, Shixiang Shane Gu, Kevin P Murphy, and Chelsea Finn. Language as an abstraction for hierarchical deep reinforcement learning. *Advances in Neural Information Processing Systems*, 32, 2019.
143. Kishor Jothimurugan, Suguman Bansal, Osbert Bastani, and Rajeev Alur. Compositional reinforcement learning from logical specifications. *Advances in Neural Information Processing Systems*, 34:10026–10039, 2021.
144. Zoe Juozapaitis, Anurag Koul, Alan Fern, Martin Erwig, and Finale Doshi-Velez. Explainable reinforcement learning via reward decomposition. In *IJCAI/ECAI Workshop on Explainable Artificial Intelligence*, 2019.
145. John Kanu, Eadom Dessalene, Xiaomin Lin, Cornelia Fermuller, and Yiannis Aloimonos. Following instructions by imagining and reaching visual goals. *arXiv preprint* arXiv:2001.09373, 2020.
146. Lena Kästner, Markus Langer, Veronika Lazar, Astrid Schomäcker, Timo Speith, and Sarah Sterz. On the relation of trust and explainability: Why to engineer for trustworthiness. In *2021 IEEE 29th International Requirements Engineering Conference Workshops (REW)*, pages 169–175. IEEE, 2021.
147. Diederik P Kingma and Max Welling. Auto-encoding variational bayes. *arXiv preprint* arXiv:1312.6114, 2013.
148. B Ravi Kiran, Ibrahim Sobh, Victor Talpaert, Patrick Mannion, Ahmad A Al Sallab, Senthil Yogamani, and Patrick Pérez. Deep reinforcement learning for autonomous driving: A survey. *IEEE Transactions on Intelligent Transportation Systems*, 2021.
149. Taisuke Kobayashi. L2c2: Locally lipschitz continuous constraint towards stable and smooth reinforcement learning. *arXiv preprint* arXiv:2202.07152, 2022.
150. Jens Kober, J Andrew Bagnell, and Jan Peters. Reinforcement learning in robotics: A survey. *The International Journal of Robotics Research*, 32(11):1238–1274, 2013.
151. Rafal Kocielnik, Saleema Amershi, and Paul N Bennett. Will you accept an imperfect ai? exploring designs for adjusting end-user expectations of ai systems. In *Proceedings of the 2019 CHI Conference on Human Factors in Computing Systems*, pages 1–14, 2019.
152. Raj Korpan and Susan L Epstein. Toward natural explanations for a robot's navigation plans. *Notes from the Explainable Robotic Systems Worshop, Human-Robot Interaction*, 2018.
153. Raj Korpan, Susan L Epstein, Anoop Aroor, and Gil Dekel. Why: Natural explanations from a robot navigator. *arXiv preprint* arXiv:1709.09741, 2017.
154. Agneza Krajna, Mario Brcic, Tomislav Lipic, and Juraj Doncevic. Explainability in reinforcement learning: perspective and position. *arXiv preprint* arXiv:2203.11547, 2022.
155. Hanna Krasowski, Xiao Wang, and Matthias Althoff. Safe reinforcement learning for autonomous lane changing using set-based prediction. In *2020 IEEE 23rd International Conference on Intelligent Transportation Systems (ITSC)*, pages 1–7. IEEE, 2020.

156. Samantha Krening and Karen M Feigh. Effect of interaction design on the human experience with interactive reinforcement learning. In *Proceedings of the 2019 on Designing Interactive Systems Conference*, pages 1089–1100, 2019.
157. Benjamin Kuipers, Edward A Feigenbaum, Peter E Hart, and Nils J Nilsson. Shakey: from conception to history. *Ai Magazine*, 38(1):88–103, 2017.
158. Anagha Kulkarni, Siddharth Srivastava, and Subbarao Kambhampati. Signaling friends and head-faking enemies simultaneously: Balancing goal obfuscation and goal legibility. *arXiv preprint* arXiv:1905.10672, 2019.
159. Karol Kurach, Anton Raichuk, Piotr Stańczyk, Michał Zając, Olivier Bachem, Lasse Espeholt, Carlos Riquelme, Damien Vincent, Marcin Michalski, Olivier Bousquet, et al. Google research football: A novel reinforcement learning environment. In *Proceedings of the AAAI Conference on Artificial Intelligence*, volume 34, pages 4501–4510, 2020.
160. Isaac Lage, Daphna Lifschitz, Finale Doshi-Velez, and Ofra Amir. Exploring computational user models for agent policy summarization. In *IJCAI: proceedings of the conference*, volume 28, page 1401. NIH Public Access, 2019.
161. Isaac Lage, Daphna Lifschitz, Finale Doshi-Velez, and Ofra Amir. Toward robust policy summarization. In *Proceedings of the 18th International Conference on Autonomous Agents and MultiAgent Systems*, pages 2081–2083, 2019.
162. Mikel Landajuela, Brenden K Petersen, Sookyung Kim, Claudio P Santiago, Ruben Glatt, Nathan Mundhenk, Jacob F Pettit, and Daniel Faissol. Discovering symbolic policies with deep reinforcement learning. In *International Conference on Machine Learning*, pages 5979–5989. PMLR, 2021.
163. Adam Daniel Laud. *Theory and application of reward shaping in reinforcement learning*. University of Illinois at Urbana-Champaign, 2004.
164. Luc Le Mero, Dewei Yi, Mehrdad Dianati, and Alexandros Mouzakitis. A survey on imitation learning techniques for end-to-end autonomous vehicles. *IEEE Transactions on Intelligent Transportation Systems*, 2022.
165. Yann LeCun, Yoshua Bengio, and Geoffrey Hinton. Deep learning. *nature*, 521(7553):436–444, 2015.
166. Jeong-Hoon Lee and Jongeun Choi. Hierarchical primitive composition: Simultaneous activation of skills with inconsistent action dimensions in multiple hierarchies. *IEEE Robotics and Automation Letters*, 2022.
167. Timothée Lesort, Natalia Díaz-Rodríguez, Jean-Franois Goudou, and David Filliat. State representation learning for control: An overview. *Neural Networks*, 108:379–392, 2018.
168. Sergey Levine, Peter Pastor, Alex Krizhevsky, Julian Ibarz, and Deirdre Quillen. Learning hand-eye coordination for robotic grasping with deep learning and large-scale data collection. *The International Journal of Robotics Research*, 37(4-5):421–436, 2018.
169. Roger J Lewis. An introduction to classification and regression tree (cart) analysis. In *Annual meeting of the society for academic emergency medicine in San Francisco, California*, volume 14, 2000.
170. Alexander C Li, Carlos Florensa, Ignasi Clavera, and Pieter Abbeel. Sub-policy adaptation for hierarchical reinforcement learning. *ICLR 2020*, 2019.
171. Guangliang Li, Randy Gomez, Keisuke Nakamura, and Bo He. Human-centered reinforcement learning: A survey. *IEEE Transactions on Human-Machine Systems*, 49(4):337–349, 2019.
172. Richard Li, Allan Jabri, Trevor Darrell, and Pulkit Agrawal. Towards practical multi-object manipulation using relational reinforcement learning. In *2020 IEEE International Conference on Robotics and Automation (ICRA)*, pages 4051–4058. IEEE, 2020.
173. Yunfei Li, Yilin Wu, Huazhe Xu, Xiaolong Wang, and Yi Wu. Solving compositional reinforcement learning problems via task reduction. *ICLR 2021*, 2021.

174. Roman Liessner, Jan Dohmen, and Marco A Wiering. Explainable reinforcement learning for longitudinal control. In *ICAART (2)*, pages 874–881, 2021.
175. Yixin Lin, Austin S Wang, Eric Undersander, and Akshara Rai. Efficient and interpretable robot manipulation with graph neural networks. *IEEE Robotics and Automation Letters*, 7(2):2740–2747, 2022.
176. Zhengxian Lin, Kim-Ho Lam, and Alan Fern. Contrastive explanations for reinforcement learning via embedded self predictions. *ICLR 2021*, 2021.
177. Martina Lippi, Petra Poklukar, Michael C Welle, Anastasiia Varava, Hang Yin, Alessandro Marino, and Danica Kragic. Latent space roadmap for visual action planning of deformable and rigid object manipulation. In *2020 IEEE/RSJ International Conference on Intelligent Robots and Systems (IROS)*, pages 5619–5626. IEEE, 2020.
178. Zachary C Lipton. The mythos of model interpretability: In machine learning, the concept of interpretability is both important and slippery. *Queue*, 16(3):31–57, 2018.
179. Guiliang Liu, Oliver Schulte, Wang Zhu, and Qingcan Li. Toward interpretable deep reinforcement learning with linear model u-trees. In *Joint European Conference on Machine Learning and Knowledge Discovery in Databases*, pages 414–429. Springer, 2018.
180. Xiao Liu, Shuyang Liu, Wenbin Li, Shangdong Yang, and Yang Gao. Keeping minimal experience to achieve efficient interpretable policy distillation. *arXiv preprint* arXiv:2203.00822, 2022.
181. Yang Liu, Shaonan Wang, Jiajun Zhang, and Chengqing Zong. Experience-based causality learning for intelligent agents. *ACM Transactions on Asian and Low-Resource Language Information Processing (TALLIP)*, 18(4):1–22, 2019.
182. Dylan P Losey, Craig G McDonald, Edoardo Battaglia, and Marcia K O'Malley. A review of intent detection, arbitration, and communication aspects of shared control for physical human–robot interaction. *Applied Mechanics Reviews*, 70(1), 2018.
183. Dylan P Losey and Marcia K O'Malley. Including uncertainty when learning from human corrections. In *Conference on Robot Learning*, pages 123–132. PMLR, 2018.
184. Jakob Løver, Vilde B Gjærum, and Anastasios M Lekkas. Explainable ai methods on a deep reinforcement learning agent for automatic docking. *IFAC-PapersOnLine*, 54(16):146–152, 2021.
185. Scott M Lundberg and Su-In Lee. A unified approach to interpreting model predictions. *Advances in neural information processing systems*, 30, 2017.
186. Jieliang Luo, Sam Green, Peter Feghali, George Legrady, and Cetin Kaya Koç. Visual diagnostics for deep reinforcement learning policy development. *arXiv preprint* arXiv:1809.06781, 2018.
187. Björn Lütjens, Michael Everett, and Jonathan P How. Safe reinforcement learning with model uncertainty estimates. In *2019 International Conference on Robotics and Automation (ICRA)*, pages 8662–8668. IEEE, 2019.
188. Prashan Madumal, Tim Miller, Liz Sonenberg, and Frank Vetere. Distal explanations for explainable reinforcement learning agents. *arXiv preprint* arXiv:2001.10284, 2020.
189. Prashan Madumal, Tim Miller, Liz Sonenberg, and Frank Vetere. Explainable reinforcement learning through a causal lens. In *Proceedings of the AAAI Conference on Artificial Intelligence*, volume 34, pages 2493–2500, 2020.
190. Francis Maes, Raphael Fonteneau, Louis Wehenkel, and Damien Ernst. Policy search in a space of simple closed-form formulas: Towards interpretability of reinforcement learning. In *International Conference on Discovery Science*, pages 37–51. Springer, 2012.
191. Luca Marzari, Ameya Pore, Diego Dall'Alba, Gerardo Aragon-Camarasa, Alessandro Farinelli, and Paolo Fiorini. Towards hierarchical task decomposition using deep reinforcement learning for pick and place subtasks. In *2021 20th International Conference on Advanced Robotics (ICAR)*, pages 640–645. IEEE, 2021.

192. Andrew Kachites McCallum et al. Learning to use selective attention and short-term memory in sequential tasks. *From animals to animats*, 4:315–324, 1996.
193. Oier Mees and Wolfram Burgard. Composing pick-and-place tasks by grounding language. In *International Symposium on Experimental Robotics*, pages 491–501. Springer, 2020.
194. John Menick. Move 37: Artificial intelligence, randomness, and creativity move 37: Artificial intelligence, randomness, and creativity. *accessed 2022-10-04*, 2016.
195. Stephanie Milani, Nicholay Topin, Manuela Veloso, and Fei Fang. A survey of explainable reinforcement learning. *arXiv preprint* arXiv:2202.08434, 2022.
196. Stephanie Milani, Zhicheng Zhang, Nicholay Topin, Zheyuan Ryan Shi, Charles Kamhoua, Evangelos E Papalexakis, and Fei Fang. Maviper: Learning decision tree policies for interpretable multi-agent reinforcement learning. *arXiv preprint* arXiv:2205.12449, 2022.
197. Aditi Mishra, Utkarsh Soni, Jinbin Huang, and Chris Bryan. Why? why not? when? visual explanations of agent behaviour in reinforcement learning. In *2022 IEEE 15th Pacific Visualization Symposium (PacificVis)*, pages 111–120. IEEE, 2022.
198. Volodymyr Mnih, Koray Kavukcuoglu, David Silver, Alex Graves, Ioannis Antonoglou, Daan Wierstra, and Martin Riedmiller. Playing atari with deep reinforcement learning. *arXiv preprint* arXiv:1312.5602, 2013.
199. Volodymyr Mnih, Koray Kavukcuoglu, David Silver, Andrei A Rusu, Joel Veness, Marc G Bellemare, Alex Graves, Martin Riedmiller, Andreas K Fidjeland, Georg Ostrovski, et al. Human-level control through deep reinforcement learning. *nature*, 518(7540):529–533, 2015.
200. T. M. Moerland, J. Broekens, and C. Jonker. Model-based reinforcement learning: A survey. *ArXiv*, abs/2006.16712, 2020.
201. Christoph Molnar. *Interpretable machine learning*. Lulu. com, 2020.
202. Grégoire Montavon, Wojciech Samek, and Klaus-Robert Müller. Methods for interpreting and understanding deep neural networks. *Digital Signal Processing*, 73:1–15, 2018.
203. Juan M Montoya, Imant Daunhawer, Julia E Vogt, and Marco A Wiering. Decoupling state representation methods from reinforcement learning in car racing. In *ICAART (2)*, pages 752–759, 2021.
204. Alexander Mott, Daniel Zoran, Mike Chrzanowski, Daan Wierstra, and Danilo Jimenez Rezende. Towards interpretable reinforcement learning using attention augmented agents. *Advances in Neural Information Processing Systems*, 32, 2019.
205. Sajad Mousavi, Michael Schukat, Enda Howley, Ali Borji, and Nasser Mozayani. Learning to predict where to look in interactive environments using deep recurrent q-learning. *arXiv preprint* arXiv:1612.05753, 2016.
206. Yazan Mualla, Igor Tchappi, Timotheus Kampik, Amro Najjar, Davide Calvaresi, Abdeljalil Abbas-Turki, Stéphane Galland, and Christophe Nicolle. The quest of parsimonious xai: A human-agent architecture for explanation formulation. *Artificial Intelligence*, 302:103573, 2022.
207. Stephen Muggleton and Luc De Raedt. Inductive logic programming: Theory and methods. *The Journal of Logic Programming*, 19:629–679, 1994.
208. Vincent C Müller. Risks of general artificial intelligence, 2014.
209. T Nathan Mundhenk, Mikel Landajuela, Ruben Glatt, Claudio P Santiago, Daniel M Faissol, and Brenden K Petersen. Symbolic regression via neural-guided genetic programming population seeding. *35th Conference on Neural Information Processing Systems (NeurIPS 2021)*, 2021.
210. Rémi Munos, Tom Stepleton, Anna Harutyunyan, and Marc Bellemare. Safe and efficient off-policy reinforcement learning. *Advances in neural information processing systems*, 29, 2016.
211. W James Murdoch, Chandan Singh, Karl Kumbier, Reza Abbasi-Asl, and Bin Yu. Interpretable machine learning: definitions, methods, and applications. *arXiv preprint* arXiv:1901.04592, 2019.

212. Ashvin Nair, Shikhar Bahl, Alexander Khazatsky, Vitchyr Pong, Glen Berseth, and Sergey Levine. Contextual imagined goals for self-supervised robotic learning. In *Conference on Robot Learning*, pages 530–539. PMLR, 2020.
213. Gabe Nelson, Aaron Saunders, and Robert Playter. The petman and atlas robots at boston dynamics. *Humanoid Robotics: A Reference*, 169:186, 2019.
214. M. Neunert, T. Boaventura, and J. Buchli. *Why Off-The-Shelf Physics Simulators Fail In Evaluating Feedback Controller Performance - A Case Study For Quadrupedal Robots: Proceedings of the 19th International Conference on CLAWAR 2016*, pages 464–472. Proceedings of the 19th International Conference on CLAWAR 2016, 10 2016.
215. Andrew Y Ng, Stuart J Russell, et al. Algorithms for inverse reinforcement learning. In *Icml*, volume 1, page 2, 2000.
216. Hai Nguyen and Hung La. Review of deep reinforcement learning for robot manipulation. In *2019 Third IEEE International Conference on Robotic Computing (IRC)*, pages 590–595. IEEE, 2019.
217. Khanh X Nguyen, Dipendra Misra, Robert Schapire, Miroslav Dudík, and Patrick Shafto. Interactive learning from activity description. In *International Conference on Machine Learning*, pages 8096–8108. PMLR, 2021.
218. Xiaotong Nie, Motoaki Hiraga, and Kazuhiro Ohkura. Visualizing deep q-learning to understanding behavior of swarm robotic system. In *Symposium on Intelligent and Evolutionary Systems*, pages 118–129. Springer, 2019.
219. Dmitry Nikulin, Anastasia Ianina, Vladimir Aliev, and Sergey Nikolenko. Free-lunch saliency via attention in atari agents. In *2019 IEEE/CVF International Conference on Computer Vision Workshop (ICCVW)*, pages 4240–4249. IEEE, 2019.
220. Justin Norden, Matthew O'Kelly, and Aman Sinha. Efficient black-box assessment of autonomous vehicle safety. *arXiv preprint* arXiv:1912.03618, 2019.
221. Ini Oguntola, Dana Hughes, and Katia Sycara. Deep interpretable models of theory of mind. In *2021 30th IEEE International Conference on Robot & Human Interactive Communication (RO-MAN)*, pages 657–664. IEEE, 2021.
222. Matthew L Olson, Roli Khanna, Lawrence Neal, Fuxin Li, and Weng-Keen Wong. Counterfactual state explanations for reinforcement learning agents via generative deep learning. *Artificial Intelligence*, 295:103455, 2021.
223. Matthew L Olson, Lawrence Neal, Fuxin Li, and Weng-Keen Wong. Counterfactual states for atari agents via generative deep learning. *arXiv preprint* arXiv:1909.12969, 2019.
224. Toby Ord. *The precipice: existential risk and the future of humanity*. Hachette Books, 2020.
225. Rohan Paleja, Yaru Niu, Andrew Silva, Chace Ritchie, Sugju Choi, and Matthew Gombolay. Learning interpretable, high-performing policies for autonomous driving. *Robotics: Science and Systems*, 2022.
226. Jun-Cheol Park, Dae-Shik Kim, and Yukie Nagai. Learning for goal-directed actions using rnnpb: Developmental change of "what to imitate". *IEEE Transactions on Cognitive and Developmental Systems*, 10(3):545–556, 2017.
227. Utsav Patel, Nithish K Sanjeev Kumar, Adarsh Jagan Sathyamoorthy, and Dinesh Manocha. Dwa-rl: Dynamically feasible deep reinforcement learning policy for robot navigation among mobile obstacles. In *2021 IEEE International Conference on Robotics and Automation (ICRA)*, pages 6057–6063. IEEE, 2021.
228. Shubham Pateria, Budhitama Subagdja, Ah-hwee Tan, and Chai Quek. Hierarchical reinforcement learning: A comprehensive survey. *ACM Computing Surveys (CSUR)*, 54(5):1–35, 2021.
229. Judea Pearl. *Causality*. Cambridge university press, 2009.
230. Judea Pearl. Causal inference. In Isabelle Guyon, Dominik Janzing, and Bernhard Schölkopf, editors, *Proceedings of Workshop on Causality: Objectives and Assessment at NIPS 2008*,

volume 6 of *Proceedings of Machine Learning Research*, pages 39–58, Whistler, Canada, 12 Dec 2010. PMLR.
231. Judea Pearl et al. Models, reasoning and inference. *Cambridge, UK: Cambridge University Press*, 19, 2000.
232. Michele Persiani and Thomas Hellström. Policy regularization for legible behavior. *arXiv preprint* arXiv:2203.04303, 2022.
233. P J Phillips, Amanda C Hahn, Peter C Fontana, David A Broniatowski, and Mark A Przybocki. Four principles of explainable artificial intelligence (draft). *National Institute of Standards and Technology (NIST)*, 2020.
234. Silviu Pitis, Elliot Creager, and Animesh Garg. Counterfactual data augmentation using locally factored dynamics. *Advances in Neural Information Processing Systems*, 33:3976–3990, 2020.
235. Athanasios S Polydoros and Lazaros Nalpantidis. Survey of model-based reinforcement learning: Applications on robotics. *Journal of Intelligent & Robotic Systems*, 86(2):153–173, 2017.
236. Hadrien Pouget, Hana Chockler, Youcheng Sun, and Daniel Kroening. Ranking policy decisions. *Advances in Neural Information Processing Systems*, 34:8702–8713, 2021.
237. Xavier Puig, Tianmin Shu, Shuang Li, Zilin Wang, Joshua B Tenenbaum, Sanja Fidler, and Antonio Torralba. Watch-and-help: A challenge for social perception and human-ai collaboration. *ICLR 2021*, 2021.
238. Erika Puiutta and Eric Veith. Explainable reinforcement learning: A survey, 2020.
239. Ahmed Hussain Qureshi, Yutaka Nakamura, Yuichiro Yoshikawa, and Hiroshi Ishiguro. Show, attend and interact: Perceivable human-robot social interaction through neural attention q-network. In *2017 IEEE International Conference on Robotics and Automation (ICRA)*, pages 1639–1645. IEEE, 2017.
240. Antonin Raffin, Ashley Hill, René Traoré, Timothée Lesort, Natalia Díaz-Rodríguez, and David Filliat. S-rl toolbox: Environments, datasets and evaluation metrics for state representation learning. *arXiv preprint* arXiv:1809.09369, 2018.
241. Antonin Raffin, Ashley Hill, René Traoré, Timothée Lesort, Natalia Díaz-Rodríguez, and David Filliat. Decoupling feature extraction from policy learning: assessing benefits of state representation learning in goal based robotics. *Workshop on "Structure and Priors in Reinforcement Learning" at ICLR 2019*, 2019.
242. Tanmay Randhavane, Aniket Bera, Emily Kubin, Austin Wang, Kurt Gray, and Dinesh Manocha. Pedestrian dominance modeling for socially-aware robot navigation. In *2019 International Conference on Robotics and Automation (ICRA)*, pages 5621–5628. IEEE, 2019.
243. Sahand Rezaei-Shoshtari, David Meger, and Inna Sharf. Learning the latent space of robot dynamics for cutting interaction inference. In *2020 IEEE/RSJ International Conference on Intelligent Robots and Systems (IROS)*, pages 5627–5632. IEEE, 2020.
244. Danilo Jimenez Rezende, Shakir Mohamed, and Daan Wierstra. Stochastic backpropagation and approximate inference in deep generative models. In *International conference on machine learning*, pages 1278–1286. PMLR, 2014.
245. Marco Tulio Ribeiro, Sameer Singh, and Carlos Guestrin. "why should i trust you?" explaining the predictions of any classifier. In *Proceedings of the 22nd ACM SIGKDD international conference on knowledge discovery and data mining*, pages 1135–1144, 2016.
246. Gaith Rjoub, Jamal Bentahar, and Omar Abdel Wahab. Explainable ai-based federated deep reinforcement learning for trusted autonomous driving. In *2022 International Wireless Communications and Mobile Computing (IWCMC)*, pages 318–323, 2022.
247. Robotnik. History of robots and robotics. *accessed 2022-10-04*, 2021.
248. Aaron M Roth. Structured representations for behaviors of autonomous robots. Master's thesis, Carnegie Mellon University, Pittsburgh, PA, July 2019.

249. Aaron M Roth, Jing Liang, and Dinesh Manocha. Xai-n: sensor-based robot navigation using expert policies and decision trees. *2021 IEEE/RSJ International Conference on Intelligent Robots and Systems (IROS)*, 2021.
250. Aaron M. Roth, Jing Liang, Ram Sriram, Elham Tabassi, and Dinesh Manocha. MSVIPER: Improved policy distillation for reinforcement-learning-based robot navigation. *Journal of the Washington Academy of Sciences*, 109(2):27–57. Summer, 2023.
251. Aaron M Roth, Samantha Reig, Umang Bhatt, Jonathan Shulgach, Tamara Amin, Afsaneh Doryab, Fei Fang, and Manuela Veloso. A robot's expressive language affects human strategy and perceptions in a competitive game. In *2019 28th IEEE International Conference on Robot and Human Interactive Communication (RO-MAN)*, pages 1–8. IEEE, 2019.
252. Aaron M Roth, Nicholay Topin, Pooyan Jamshidi, and Manuela Veloso. Conservative q-improvement: Reinforcement learning for an interpretable decision-tree policy. *arXiv preprint* arXiv:1907.01180, 2019.
253. Cynthia Rudin. Stop explaining black box machine learning models for high stakes decisions and use interpretable models instead. *Nature Machine Intelligence*, 1(5):206–215, 2019.
254. Cynthia Rudin, Chaofan Chen, Zhi Chen, Haiyang Huang, Lesia Semenova, and Chudi Zhong. Interpretable machine learning: Fundamental principles and 10 grand challenges. *Statistics Surveys*, 16:1–85, 2022.
255. Cynthia Rudin and Joanna Radin. Why are we using black box models in ai when we don't need to? a lesson from an explainable ai competition. *Harvard Data Science Review*, 1(2), 2019.
256. Andrei A Rusu, Sergio Gomez Colmenarejo, Caglar Gulcehre, Guillaume Desjardins, James Kirkpatrick, Razvan Pascanu, Volodymyr Mnih, Koray Kavukcuoglu, and Raia Hadsell. Policy distillation. *arXiv preprint* arXiv:1511.06295, 2015.
257. Fatai Sado, Chu Kiong Loo, Matthias Kerzel, and Stefan Wermter. Explainable goal-driven agents and robots – a comprehensive review and new framework, 2020.
258. Tatsuya Sakai and Takayuki Nagai. Explainable autonomous robots: a survey and perspective. *Advanced Robotics*, 36(5-6):219–238, 2022.
259. Ahmad EL Sallab, Mohammed Abdou, Etienne Perot, and Senthil Yogamani. Deep reinforcement learning framework for autonomous driving. *Electronic Imaging*, 2017(19):70–76, 2017.
260. Lindsay Sanneman and Julie Shah. Explaining reward functions to humans for better human-robot collaboration. *Association for the Advancement of Artificial Intelligence*, 2021.
261. Kristin E Schaefer, Edward R Straub, Jessie YC Chen, Joe Putney, and Arthur W Evans III. Communicating intent to develop shared situation awareness and engender trust in human-agent teams. *Cognitive Systems Research*, 46:26–39, 2017.
262. John Schulman, Philipp Moritz, Sergey Levine, Michael Jordan, and Pieter Abbeel. High-dimensional continuous control using generalized advantage estimation. *ICLR 2016*, 2015.
263. Alessandra Sciutti, Martina Mara, Vincenzo Tagliasco, and Giulio Sandini. Humanizing human-robot interaction: On the importance of mutual understanding. *IEEE Technology and Society Magazine*, 37(1):22–29, 2018.
264. Maximilian Seitzer, Bernhard Schölkopf, and Georg Martius. Causal influence detection for improving efficiency in reinforcement learning. *Advances in Neural Information Processing Systems*, 34:22905–22918, 2021.
265. Pedro Sequeira and Melinda Gervasio. Interestingness elements for explainable reinforcement learning: Understanding agents' capabilities and limitations. *Artificial Intelligence*, 288:103367, 2020.
266. Pedro Sequeira, Eric Yeh, and Melinda T Gervasio. Interestingness elements for explainable reinforcement learning through introspection. In *IUI Workshops*, page 7, 2019.
267. Wenjie Shang, Qingyang Li, Zhiwei Qin, Yang Yu, Yiping Meng, and Jieping Ye. Partially observable environment estimation with uplift inference for reinforcement learning based recommendation. *Machine Learning*, pages 1–38, 2021.

268. Lin Shao, Toki Migimatsu, Qiang Zhang, Karen Yang, and Jeannette Bohg. Concept2robot: Learning manipulation concepts from instructions and human demonstrations. In *Proceedings of Robotics: Science and Systems (RSS)*, 2020.
269. Hassam Sheikh, Shauharda Khadka, Santiago Miret, and Somdeb Majumdar. Learning intrinsic symbolic rewards in reinforcement learning. *arXiv preprint* arXiv:2010.03694, 2020.
270. Wenjie Shi, Gao Huang, Shiji Song, Zhuoyuan Wang, Tingyu Lin, and Cheng Wu. Self-supervised discovering of interpretable features for reinforcement learning. *IEEE Transactions on Pattern Analysis and Machine Intelligence*, 2020.
271. Avanti Shrikumar, Peyton Greenside, and Anshul Kundaje. Learning important features through propagating activation differences. In *International Conference on Machine Learning*, pages 3145–3153. PMLR, 2017.
272. Tianmin Shu, Abhishek Bhandwaldar, Chuang Gan, Kevin Smith, Shari Liu, Dan Gutfreund, Elizabeth Spelke, Joshua Tenenbaum, and Tomer Ullman. Agent: A benchmark for core psychological reasoning. In *International Conference on Machine Learning*, pages 9614–9625. PMLR, 2021.
273. Tianmin Shu, Caiming Xiong, and Richard Socher. Hierarchical and interpretable skill acquisition in multi-task reinforcement learning. *arXiv preprint* arXiv:1712.07294, 2017.
274. Maximilian Sieb, Zhou Xian, Audrey Huang, Oliver Kroemer, and Katerina Fragkiadaki. Graph-structured visual imitation. In *Conference on Robot Learning*, pages 979–989. PMLR, 2020.
275. Joseph Sifakis. Autonomous systems–an architectural characterization. In *Models, Languages, and Tools for Concurrent and Distributed Programming*, pages 388–410. Springer, 2019.
276. Andrew Silva and Matthew Gombolay. Empirically evaluating meta learning of robot explainability with humans. *under submission*, 2022.
277. Andrew Silva, Matthew Gombolay, Taylor Killian, Ivan Jimenez, and Sung-Hyun Son. Optimization methods for interpretable differentiable decision trees applied to reinforcement learning. In *International Conference on Artificial Intelligence and Statistics*, pages 1855–1865. PMLR, 2020.
278. David Silver, Aja Huang, Chris J Maddison, Arthur Guez, Laurent Sifre, George Van Den Driessche, Julian Schrittwieser, Ioannis Antonoglou, Veda Panneershelvam, Marc Lanctot, et al. Mastering the game of go with deep neural networks and tree search. *nature*, 529(7587):484–489, 2016.
279. Tom Silver, Ashay Athalye, Joshua B. Tenenbaum, Tomas Lozano-Perez, and Leslie Pack Kaelbling. Learning neuro-symbolic skills for bilevel planning, 2022.
280. Karen Simonyan, Andrea Vedaldi, and Andrew Zisserman. Deep inside convolutional networks: Visualising image classification models and saliency maps. *arXiv preprint* arXiv:1312.6034, 2013.
281. Dan Siroker and Pete Koomen. *A/B testing: The most powerful way to turn clicks into customers*. John Wiley & Sons, 2013.
282. Shagun Sodhani, Amy Zhang, and Joelle Pineau. Multi-task reinforcement learning with context-based representations. In *International Conference on Machine Learning*, pages 9767–9779. PMLR, 2021.
283. Philippe Sormani. Logic-in-action? alphago, surprise move 37 and interaction analysis. In *Handbook of the 6th World Congress and School on Universal Logic*, page 378, 2018.
284. Ivan Sorokin, Alexey Seleznev, Mikhail Pavlov, Aleksandr Fedorov, and Anastasiia Ignateva. Deep attention recurrent q-network. *arXiv preprint* arXiv:1512.01693, 2015.
285. Sarath Sreedharan, Subbarao Kambhampati, et al. Handling model uncertainty and multiplicity in explanations via model reconciliation. In *Proceedings of the International Conference on Automated Planning and Scheduling*, volume 28, 2018.

286. Sarath Sreedharan, Utkarsh Soni, Mudit Verma, Siddharth Srivastava, and Subbarao Kambhampati. Bridging the gap: Providing post-hoc symbolic explanations for sequential decision-making problems with inscrutable representations, 2020.
287. Oliver Struckmeier, Mattia Racca, and Ville Kyrki. Autonomous generation of robust and focused explanations for robot policies. In *2019 28th IEEE International Conference on Robot and Human Interactive Communication (RO-MAN)*, pages 1–8. IEEE, 2019.
288. Roykrong Sukkerd, Reid Simmons, and David Garlan. Tradeoff-focused contrastive explanation for mdp planning. In *2020 29th IEEE International Conference on Robot and Human Interactive Communication (RO-MAN)*, pages 1041–1048. IEEE, 2020.
289. Theodore R Sumers, Robert D Hawkins, Mark K Ho, Thomas L Griffiths, and Dylan Hadfield-Menell. How to talk so your robot will learn: Instructions, descriptions, and pragmatics. *arXiv preprint* arXiv:2206.07870, 2022.
290. Richard S Sutton and Andrew G Barto. *Reinforcement learning: An introduction*. MIT press, 2018.
291. Richard S Sutton, Joseph Modayil, Michael Delp, Thomas Degris, Patrick M Pilarski, Adam White, and Doina Precup. Horde: A scalable real-time architecture for learning knowledge from unsupervised sensorimotor interaction. In *The 10th International Conference on Autonomous Agents and Multiagent Systems-Volume 2*, pages 761–768, 2011.
292. Aaquib Tabrez and Bradley Hayes. Improving human-robot interaction through explainable reinforcement learning. In *2019 14th ACM/IEEE International Conference on Human-Robot Interaction (HRI)*, pages 751–753. IEEE, 2019.
293. Aaquib Tabrez, Matthew B Luebbers, and Bradley Hayes. Automated failure-mode clustering and labeling for informed car-to-driver handover in autonomous vehicles. *HRI '20 Workshop on Assessing, Explaining, and Conveying Robot Proficiency for Human-Robot Teaming*, 2020.
294. Lei Tai, Peng Yun, Yuying Chen, Congcong Liu, Haoyang Ye, and Ming Liu. Visual-based autonomous driving deployment from a stochastic and uncertainty-aware perspective. In *2019 IEEE/RSJ International Conference on Intelligent Robots and Systems (IROS)*, pages 2622–2628. IEEE, 2019.
295. Victor Talpaert, Ibrahim Sobh, B Ravi Kiran, Patrick Mannion, Senthil Yogamani, Ahmad El-Sallab, and Patrick Perez. Exploring applications of deep reinforcement learning for real-world autonomous driving systems. *arXiv preprint* arXiv:1901.01536, 2019.
296. Diane Tang, Ashish Agarwal, Deirdre O'Brien, and Mike Meyer. Overlapping experiment infrastructure: More, better, faster experimentation. In *Proceedings of the 16th ACM SIGKDD international conference on Knowledge discovery and data mining*, pages 17–26, 2010.
297. Matthew E Taylor, Nicholas Nissen, Yuan Wang, and Neda Navidi. Improving reinforcement learning with human assistance: an argument for human subject studies with hippo gym. *Neural Computing and Applications*, pages 1–11, 2021.
298. Sravanthi Thota, R Nethravathi, S Naresh Kumar, and M Shyamsunder. An analysis of reinforcement learning interpretation techniques. In *AIP Conference Proceedings*, volume 2418, page 020050. AIP Publishing LLC, 2022.
299. Jakob Thumm and Matthias Althoff. Provably safe deep reinforcement learning for robotic manipulation in human environments. *arXiv preprint* arXiv:2205.06311, 2022.
300. Emanuel Todorov, Tom Erez, and Yuval Tassa. Mujoco: A physics engine for model-based control. In *2012 IEEE/RSJ International Conference on Intelligent Robots and Systems*, pages 5026–5033. IEEE, 2012.
301. Nicholay Topin, Stephanie Milani, Fei Fang, and Manuela Veloso. Iterative bounding mdps: Learning interpretable policies via non-interpretable methods. In *Proceedings of the AAAI Conference on Artificial Intelligence*, volume 35, pages 9923–9931, 2021.

302. Nicholay Topin and Manuela Veloso. Generation of policy-level explanations for reinforcement learning. In *Proceedings of the AAAI Conference on Artificial Intelligence*, volume 33, pages 2514–2521, 2019.
303. Wolfgang Trautwein and G Dean Robinson. Operational loopwheel suspension system for mars rover demonstration model. Technical report, Lockheed Missiles and Space Co., 1978.
304. Silvia Tulli, Marta Couto, Miguel Vasco, Elmira Yadollahi, Francisco Melo, and Ana Paiva. Explainable agency by revealing suboptimality in child-robot learning scenarios. In *International Conference on Social Robotics*, pages 23–35. Springer, 2020.
305. Jeffrey M Turner. A propulsion and steering control system for the mars rover. Technical report, Renssalaer Polytechnic Institute, 1980.
306. William TB Uther and Manuela M Veloso. Tree based discretization for continuous state space reinforcement learning. *Aaai/iaai*, 98:769–774, 1998.
307. Laurens Van der Maaten and Geoffrey Hinton. Visualizing data using t-sne. *Journal of machine learning research*, 9(11), 2008.
308. Sanne van Waveren, Christian Pek, Jana Tumova, and Iolanda Leite. Correct me if i'm wrong: Using non-experts to repair reinforcement learning policies. In *Proceedings of the 17th ACM/IEEE International Conference on Human-Robot Interaction*, pages 1–9, 2022.
309. Marko Vasic, Andrija Petrovic, Kaiyuan Wang, Mladen Nikolic, Rishabh Singh, and Sarfraz Khurshid. Moët: Interpretable and verifiable reinforcement learning via mixture of expert trees. *arXiv preprint* arXiv:1906.06717, 2019.
310. Sagar Gubbi Venkatesh, Raviteja Upadrashta, and Bharadwaj Amrutur. Translating natural language instructions to computer programs for robot manipulation. In *2021 IEEE/RSJ International Conference on Intelligent Robots and Systems (IROS)*, pages 1919–1926. IEEE, 2021.
311. Frank MF Verberne, Jaap Ham, and Cees JH Midden. Trust in smart systems: Sharing driving goals and giving information to increase trustworthiness and acceptability of smart systems in cars. *Human factors*, 54(5):799–810, 2012.
312. Abhinav Verma. Verifiable and interpretable reinforcement learning through program synthesis. In *Proceedings of the AAAI Conference on Artificial Intelligence*, volume 33, pages 9902–9903, 2019.
313. Abhinav Verma, Vijayaraghavan Murali, Rishabh Singh, Pushmeet Kohli, and Swarat Chaudhuri. Programmatically interpretable reinforcement learning. In *International Conference on Machine Learning*, pages 5045–5054. PMLR, 2018.
314. Mathurin Videau, Alessandro Leite, Olivier Teytaud, and Marc Schoenauer. Multi-objective genetic programming for explainable reinforcement learning. In *European Conference on Genetic Programming (Part of EvoStar)*, pages 278–293. Springer, 2022.
315. Amelec Viloria, Nelson Alberto Lizardo Zelaya, and Noel Varela. Design and simulation of vehicle controllers through genetic algorithms. *Procedia Computer Science*, 175:453–458, 2020.
316. Paul Voigt and Axel Von dem Bussche. The eu general data protection regulation (gdpr). *A Practical Guide, 1st Ed., Cham: Springer International Publishing*, 10:3152676, 2017.
317. Sergei Volodin, Nevan Wichers, and Jeremy Nixon. Resolving spurious correlations in causal models of environments via interventions. *arXiv preprint* arXiv:2002.05217, 2020.
318. George A Vouros. Explainable deep reinforcement learning: State of the art and challenges. *ACM Computing Surveys (CSUR)*, 2022.
319. Michael Walker, Hooman Hedayati, Jennifer Lee, and Daniel Szafir. Communicating robot motion intent with augmented reality. In *Proceedings of the 2018 ACM/IEEE International Conference on Human-Robot Interaction*, pages 316–324, 2018.
320. Sebastian Wallkotter, Mohamed Chetouani, and Ginevra Castellano. A new approach to evaluating legibility: Comparing legibility frameworks using framework-independent robot motion trajectories. *arXiv preprint* arXiv:2201.05765, 2022.

321. Jane Wang, Michael King, Nicolas Porcel, Zeb Kurth-Nelson, Tina Zhu, Charlie Deck, Peter Choy, Mary Cassin, Malcolm Reynolds, Francis Song, Gavin Buttimore, David Reichert, Neil Rabinowitz, Loic Matthey, Demis Hassabis, Alex Lerchner, and Matthew Botvinick. Alchemy: A structured task distribution for meta-reinforcement learning. *arXiv preprint* arXiv:2102.02926, 2021.
322. Ning Wang, David V Pynadath, and Susan G Hill. Trust calibration within a human-robot team: Comparing automatically generated explanations. In *2016 11th ACM/IEEE International Conference on Human-Robot Interaction (HRI)*, pages 109–116. IEEE, 2016.
323. Ziyu Wang, Tom Schaul, Matteo Hessel, Hado Hasselt, Marc Lanctot, and Nando Freitas. Dueling network architectures for deep reinforcement learning. In *International conference on machine learning*, pages 1995–2003. PMLR, 2016.
324. Katharina Weitz. Towards human-centered ai: Psychological concepts as foundation for empirical xai research. *it-Information Technology*, 64(1-2):71–75, 2022.
325. Adrian Weller. Transparency: motivations and challenges. In *Explainable AI: Interpreting, Explaining and Visualizing Deep Learning*, pages 23–40. Springer, 2019.
326. Lindsay Wells and Tomasz Bednarz. Explainable ai and reinforcement learning-a systematic review of current approaches and trends. *Frontiers in artificial intelligence*, 4:48, 2021.
327. Erik Wijmans, Julian Straub, Dhruv Batra, Irfan Essa, Judy Hoffman, and Ari Morcos. Analyzing visual representations in embodied navigation tasks. *arXiv preprint* arXiv:2003.05993, 2020.
328. MJ Willis. Proportional-integral-derivative control. *Dept. of Chemical and Process Engineering University of Newcastle*, 1999.
329. Christian Wirth, Riad Akrour, Gerhard Neumann, Johannes Fürnkranz, et al. A survey of preference-based reinforcement learning methods. *Journal of Machine Learning Research*, 18(136):1–46, 2017.
330. Robert H Wortham, Andreas Theodorou, and Joanna J Bryson. Improving robot transparency: Real-time visualisation of robot ai substantially improves understanding in naive observers. In *2017 26th IEEE international symposium on robot and human interactive communication (RO-MAN)*, pages 1424–1431. IEEE, 2017.
331. Jinwei Xing, Takashi Nagata, Xinyun Zou, Emre Neftci, and Jeffrey L Krichmar. Policy distillation with selective input gradient regularization for efficient interpretability. *arXiv preprint* arXiv:2205.08685, 2022.
332. Fangzhou Xiong, Zhiyong Liu, Kaizhu Huang, Xu Yang, and Hong Qiao. State primitive learning to overcome catastrophic forgetting in robotics. *Cognitive Computation*, 13(2):394–402, 2021.
333. Fangzhou Xiong, Zhiyong Liu, Kaizhu Huang, Xu Yang, Hong Qiao, and Amir Hussain. Encoding primitives generation policy learning for robotic arm to overcome catastrophic forgetting in sequential multi-tasks learning. *Neural Networks*, 129:163–173, 2020.
334. Duo Xu and Faramarz Fekri. Interpretable model-based hierarchical reinforcement learning using inductive logic programming. Technical report, EasyChair, 2021.
335. Tsung-Yen Yang, Tingnan Zhang, Linda Luu, Sehoon Ha, Jie Tan, and Wenhao Yu. Safe reinforcement learning for legged locomotion. *arXiv preprint* arXiv:2203.02638, 2022.
336. Zhao Yang, Song Bai, Li Zhang, and Philip HS Torr. Learn to interpret atari agents. *arXiv preprint* arXiv:1812.11276, 2018.
337. Rex Ying, Dylan Bourgeois, Jiaxuan You, Marinka Zitnik, and Jure Leskovec. Gnnexplainer: Generating explanations for graph neural networks. *Advances in neural information processing systems*, 32:9240, 2019.
338. Tianhe Yu, Deirdre Quillen, Zhanpeng He, Ryan Julian, Karol Hausman, Chelsea Finn, and Sergey Levine. Meta-world: A benchmark and evaluation for multi-task and meta reinforcement learning. In *Conference on Robot Learning*, pages 1094–1100. PMLR, 2020.

339. Wenhao Yu, Visak CV Kumar, Greg Turk, and C Karen Liu. Sim-to-real transfer for biped locomotion. In *2019 IEEE/RSJ International Conference on Intelligent Robots and Systems (IROS)*, pages 3503–3510. IEEE, 2019.
340. Liu Yuezhang, Ruohan Zhang, and Dana H Ballard. An initial attempt of combining visual selective attention with deep reinforcement learning. *arXiv preprint* arXiv:1811.04407, 2018.
341. Tom Zahavy, Nir Ben-Zrihem, and Shie Mannor. Graying the black box: Understanding dqns. In *International Conference on Machine Learning*, pages 1899–1908. PMLR, 2016.
342. Mehrdad Zakershahrak and Samira Ghodratnama. Are we on the same page? hierarchical explanation generation for planning tasks in human-robot teaming using reinforcement learning. *arXiv preprint* arXiv:2012.11792, 2020.
343. Vinicius Zambaldi, David Raposo, Adam Santoro, Victor Bapst, Yujia Li, Igor Babuschkin, Karl Tuyls, David Reichert, Timothy Lillicrap, Edward Lockhart, et al. Relational deep reinforcement learning. *arXiv preprint* arXiv:1806.01830, 2018.
344. Amber E Zelvelder, Marcus Westberg, and Kary Främling. Assessing explainability in reinforcement learning. In *International Workshop on Explainable, Transparent Autonomous Agents and Multi-Agent Systems*, pages 223–240. Springer, 2021.
345. Haodi Zhang, Zihang Gao, Yi Zhou, Hao Zhang, Kaishun Wu, and Fangzhen Lin. Faster and safer training by embedding high-level knowledge into deep reinforcement learning. *arXiv preprint* arXiv:1910.09986, 2019.
346. Jesse Zhang, Haonan Yu, and Wei Xu. Hierarchical reinforcement learning by discovering intrinsic options. *ICLR 2021*, 2021.
347. Linrui Zhang, Qin Zhang, Li Shen, Bo Yuan, and Xueqian Wang. Saferl-kit: Evaluating efficient reinforcement learning methods for safe autonomous driving. *Workshop on Safe Learning for Autonomous Driving (SL4AD) in the 39 th International Conference on Machine Learning (ICML)*, 2022.
348. Marvin Zhang, Sharad Vikram, Laura Smith, Pieter Abbeel, Matthew Johnson, and Sergey Levine. Solar: Deep structured representations for model-based reinforcement learning. In *International Conference on Machine Learning*, pages 7444–7453. PMLR, 2019.
349. Ruohan Zhang, Sihang Guo, Bo Liu, Yifeng Zhu, Mary Hayhoe, Dana Ballard, and Peter Stone. Machine versus human attention in deep reinforcement learning tasks. *arXiv preprint* arXiv:2010.15942, 2020.
350. Tony Z Zhao, Anusha Nagabandi, Kate Rakelly, Chelsea Finn, and Sergey Levine. Meld: Meta-reinforcement learning from images via latent state models. *4th Conference on Robot Learning (CoRL 2020), Cambridge MA, USA.*, 2020.
351. Wenshuai Zhao, Jorge Peña Queralta, and Tomi Westerlund. Sim-to-real transfer in deep reinforcement learning for robotics: a survey. In *2020 IEEE Symposium Series on Computational Intelligence (SSCI)*, pages 737–744. IEEE, 2020.
352. Xuan Zhao, Tingxiang Fan, Dawei Wang, Zhe Hu, Tao Han, and Jia Pan. An actor-critic approach for legible robot motion planner. In *2020 IEEE International Conference on Robotics and Automation (ICRA)*, pages 5949–5955. IEEE, 2020.
353. Lei Zheng, Siu-Yeung Cho, and Chai Quek. Reinforcement based u-tree: A novel approach for solving pomdp. In *Handbook on Decision Making*, pages 205–232. Springer, 2010.
354. Yilun Zhou, Serena Booth, Nadia Figueroa, and Julie Shah. Rocus: Robot controller understanding via sampling. In *Conference on Robot Learning*, pages 850–860. PMLR, 2022.
355. Zhuangdi Zhu, Kaixiang Lin, and Jiayu Zhou. Transfer learning in deep reinforcement learning: A survey. *arXiv preprint* arXiv:2009.07888, 2020.
356. Nir Ben Zrihem, Tom Zahavy, and Shie Mannor. Visualizing dynamics: from t-sne to semi-mdps. *2016 ICML Workshop on Human Interpretability in Machine Learning (WHI 2016)*, 2016.

Index

A
Actor critic (A2C), 34, 78
Adversarial attacks, 12
Alchemy, 75
Analyze interaction
   analyze interaction for goal understanding, 26, 29, 46
   analyze interaction for policy understanding, 26, 29, 45, 61
   analyze training interaction for transition model, 26, 29, 47–48
Applications
   autonomous vehicles, 6, 8, 31, 35, 43, 46, 48, 51, 53, 73, 81, 82, 84
   drones, 33, 38
   finance, 8
   healthcare, 8, 16
   law, 8, 82
   logistics, 87
   manipulation, 6, 52, 54, 59, 65, 72
   ocean robotics, 31
   physical component quality control, 56
   ridesharing, 82
   search and rescue, 16, 55, 73, 82
   smart home, 58, 83
   social robotics, 35, 64, 74, 76
   video games, 36–38, 53, 73, 81
Atari, 34, 36, 40, 41, 43, 53, 78, 81
Attention, 42, 71, 79
Audience, 5, 12, 16, 17, 20, 69, 73, 78, 88

B
Barrett Electronics Corporation, 1
Baxter robot, 42
Bayesian optimization, 38
Bipedal walker robot, 47
Black box, 2, 3, 8, 30, 31, 45, 52, 54, 71, 77, 80, 83
Boston Dynamics, 1

C
CARLA autonomous driving simulator, 35, 51, 53
Causal DT explanation, 24, 27, 30, 32
Causal influence models, 26, 29, 58, 59
Causal methods, 11–13
   structured causal models (SCM), 12
Causal world, 76
Causality, 31, 37, 47, 58–59, 69, 71, 79, 85–86
Certain model reconciliation, 25, 28, 57, 85
Classification and regression tree (CART), 31
Classification system, 4–6, 8, 11–21
Clustering, 39–41, 43, 80
Conservative Q-improvement, 19, 30, 49, 75, 77, 78, 80
Constrained execution, 50, 83
Constrained learning, 44, 50, 83
Constrained policy optimization (CPO), 50
Contrastive explanations, 36, 38, 73
Contrastive Explanations as Justification, 37

Contrastive explanations as justification, 26, 29, 37, 38
Contrastive explanations learned during training, 25, 28, 36–37
Counterfactual explanations (GAN), 25, 26, 28, 29, 31, 36–38, 58, 61, 71, 75, 79–80, 82, 86
Counterfactual explanations (non causal), 19
Counterfactual explanations (SCM), 26, 28, 37–38, 58–59, 79
Critical states, 38, 45, 46, 71, 81
Custom domain specific language as action components, 27, 83

D
Debugging, 3, 61, 71, 73
Decision tree, 18, 19, 23–32, 61, 75, 77–78, 80, 81, 87
Deep Q-network (DQN), 34, 35, 40, 50, 78
Defense advanced research projects agency (DARPA), 3
Demonstration, learning from, 42, 43, 62, 64, 66
Differential decision tree policy, 24, 27, 30–32, 77
Dimension reduction, 36, 39, 86
Domain knowledge, 42, 49, 52, 55
Dynamic window approach for reinforcement learning (DWA-RL), 51
Dynamics, 43

E
End-user, 5, 15, 17, 20, 72, 73, 83
Ethics, 8
Evolutionary algorithms, 30, 55, 84, 87
Existential risk, 75
Expert policy, 30–31
Explainability, definition of, 3, 11, 13, 15
Explainable artificial intelligence (XAI), 4, 8, 12
Explainable reinforcement learning (XRL), 8–11
Explainable robotics (X-robotics), 8, 12–16
Explanation accuracy, 4, 12, 20, 27, 71, 72
  certain, 17, 29, 30, 32, 40, 52, 54–59, 64
  uncertain, measured, 17, 29, 49
  uncertain, unmeasured, 17, 23, 29
Explanation, definition of, 3

Explanation format, 12

F
Fetch3D robot arm, 44, 52
Fidelity, 12
Format, 4, 13, 16, 17, 19–20, 27, 29, 69
  causal model, 58
  feature weights/importance, 47, 70
  hidden policy variables, 47
  high-level action/skill selection, 54, 64, 66
  instructions, 62, 72
  language or text, 49, 54–57, 60, 62, 84
  model uncertainty, 50
  pixels (individual features), 33–35
  representation (transformed state), 40, 50, 65
  reward weights, 59, 60, 69, 74
  rules or constraints, 49
  simplified/abstract state space, 39
  state examples (images or text description), 36, 37, 58
  state examples (trajectories, videos), 45, 46, 48
  symbolic structure, 64, 65, 87
  task plan, 52, 65, 80
  tree, 30, 32, 58, 75, 77, 87
  trees, mixture of, 32, 75, 78

G
General adversarial network (GAN), 37, 79
General artificial intelligence, 75
General data protection regulation (GDPR), 3, 8
General Motors, 1
Generalized value functions, 37
Genetic algorithms, 30, 55, 84, 87
Global summaries, 33, 82
Google research football environment, 65
Graph, 41, 42, 51, 53, 62, 65–66, 79–81, 85, 86
Gridworld, 38, 50, 56, 80, 85

H
Hard attributes, 16–20
Hierarchical reinforcement learning
  primitive generation, 11
  skills or goals, 12
Hierarchical RL, 39
  high level interpretability via hierarchical RL, 25, 28, 52–53, 84

Index

primitive or skill generation for hierarchical RL, 25, 28, 39, 54, 84
Hindsight experience replay, 63
HIPPO Gym, 76
Human-in-the-loop, 12
Human-in-the-loop correction, 24, 27, 62–64, 79
Human-robot interaction (HRI), 34–35, 43, 47, 50, 51, 55, 60, 64, 70, 73–74, 76, 78, 83, 84, 87, 88

I

Imitation learning (IL), 7, 12
Improving legibility, 24, 27, 66–67
Inductive logic, 42
Input perturbation, 35–37
Instruction following, 11, 19, 24, 27, 54, 62–64, 86–87
Intepretability, definition of, 3, 14
Interactive transferable augmented instruction graphs (ITAIG), 62
Interestingness, 12
Interpretation, definition of, 3
Inverse reinforcement learning, 74

K

Knowledge limits, 4, 12, 14, 20, 27, 50, 55, 57, 72, 82, 83, 85, 86
   complete, 17, 29, 30, 32, 56
   not attempted, 17, 23, 29, 62
   partially attempted, 17, 29, 47, 50, 62
Knowledge representation, 2

L

Latent spaces or latent states, 28, 41–44, 50–52, 62, 75, 81
Learn DT by policy distillation, 24, 27, 30–32
Learning from demonstration, 42, 43, 62, 64, 66
Learn mixture of DT by policy distillation, 23, 27, 32
Learn soft decision tree by policy distillation, 24, 27, 32, 75
Legibility, 5, 12, 14, 15, 21, 66–67, 74, 82, 87
   attempted, 17, 34, 66
   method, 11, 66–67
   not attempted, 17, 23

Level of detail, 15
Level of evaluation, 15
Linear model U-trees, 12

M

Manipulation, 36, 41, 42, 44, 50, 52, 54, 59, 65–67, 72
Markov decision process (MDP), 19, 20, 36–41, 52, 53, 55, 56, 61, 72, 77, 79, 80
Marvin Minsky, 1
Meaningful representation learning, 19, 39, 60, 70, 80–81
   autonomously discovered latent states, 25, 28, 40–44
   graph representations of visual input, 25, 28, 40–44, 65–66
   learning logical relationships, 25, 28, 40–44, 65–66
   metadata or human/domain data, 25, 28, 40–44
Memory-based probability-of-success, 20, 26, 29, 60
Meta world, 43, 75
Minimally complete explanation, 37
Minimal sufficient explanation (w/reward decomposition), 26, 29, 59
Model-agnostic or model-specific (MAMS), 4, 13, 17, 23
   model-agnostic, 17, 24, 26, 35, 39, 40, 44–47, 60, 64–66
   model-specific, 17, 24, 26, 30, 32–34, 36, 37, 40, 45, 46, 48–50, 52, 54–60, 62, 65
   variable, 17, 24, 26, 40, 47
Model checking, 37, 79–80
Model reconciliation, 19
Model uncertainty, 12
MuJoCo, 52, 65
Multi-agent, 31, 87
Multiple scenario verifiable reinforcement learning via policy extraction (MSVIPER), 19, 24, 27, 30, 31, 49, 75, 78, 80, 83, 84

N

National Aeronautics and Space Administration (NASA), 1
National institute for standards and technology (NIST), 8, 12, 14, 20

Natural language, 7, 20, 54–56, 62, 63, 70, 72, 73, 84–87
Natural language processing (NLP), 7
Navigation, robot, 31, 36, 51, 57, 64, 67
Neural nets, 2, 30–34, 42, 49, 54, 65, 66, 77–79, 86
   convolutional neural nets, 33
   graph neural nets, 42, 66
   recurrent neural network, 65
NLP template for model or policy, 26, 29, 55, 75, 84, 85
NLP templates for queries, 26, 29, 56, 85
Nonlinear decision trees (NLDT), 31, 32

O

Observation analysis, 12, 13, 18, 61, 71, 81–82
   A/B testing, 11, 26, 29, 47
   interrogative observation analysis, 26, 29, 48
   statistical or frequency techniques, 12, 19
   training Data, 11
Observation and network analysis: SHAP feature attribution, 26, 29, 47–48
OpenAI gym, 43, 76

P

Partially observable Markov decision process (POMDP), 55, 84
Pepper robot, 35, 74
Performance tradeoff, 8
Physical world, 9
Planning-domain definition language (PDDL), 57
Policy distillation, 19, 30, 35, 49, 64–65, 78
Predictability, 5, 12, 14, 15, 21
   attempted, 17, 46
   not attempted, 17, 23
Primitive generation, 11
Primitive or skill generation for hierarchical RL, 25, 28, 54, 84
Programmatically interpretable reinforcement learning (PIRL), 12, 49, 83
Proportional-integral-derivative (PID) control, 44, 81

Q

Q-network, 33, 34

Query-based, 12

R

Reactivity, 5, 12, 15, 17, 21
Readability, 5, 12, 14, 15, 21, 66, 74, 82, 87
   attempted, 17, 34
   not attempted, 17, 23
Real world, 8, 9, 36
Recurrent attention model, 34
Regulations, 8
Relational learning, 42
Representation learning, 12, 18, 40
Rethink robotics Sawyer robot, 60
Reward augmentation and repair, 26, 29, 60, 85
Reward decomposition, 11, 12, 18, 26, 29, 59–60, 69, 74, 85–86
Reward shaping, 74
RL visualization for debugging, 24, 27, 61
Robot navigation, 31, 36, 51, 57, 64, 67
Rule-interposing learning (RIL), 50

S

SafeRL-Kit, 75
Safety, 33, 46, 49–51, 55, 73–75, 78, 81–83, 86
   physical, 3
   regulations, 8
   safe RL, 3, 12
Safety-constrained learning, 25, 28, 50, 75, 83
Safety in execution: safety via planning, 25, 28, 40, 50–51, 75
Safety-informed execution: uncertainty aware action-selection, 25, 28, 50–51, 75, 83
Saliency visualization, 12, 18–20, 33–36, 45, 69, 71, 78–79, 81
   as attention side effect, 24, 27, 34–35, 61
   by forward propagation or backpropagation, 24, 27, 33–35, 75
   by input perturbation, 18, 24, 27, 35–36, 61
Scope, 4, 12, 13, 18, 23
   either, 17, 24, 26, 40, 65
   global, 17, 24, 26, 33, 39, 40, 45–47, 49, 50, 52, 58, 64–66
   local, 17, 24, 26, 33–37, 40, 47, 48, 50, 54–60, 62, 65
Self-explainable or post-hoc (SEPH), 4, 12, 13, 18–19, 23

post-hoc, 17, 24, 26, 33, 35–37, 39, 45–48, 55–60, 62, 64
self-explainable, 17, 24, 26, 30, 32, 34, 39, 40, 49, 50, 52, 54, 56, 58, 62, 65, 66
Self-supervised learning, 43
Semi Markov decision process (SMDP), 40
Sensor Based Navigation (XAI-N), 30, 31, 49, 75, 80, 83, 84
Shadow hand, 52
SHAKEY robot, 1
SHAP feature attribution, 29, 47–48
SHapley additive exPlanations (SHAP), 48
Simulation, 9
Soft attributes, 16
Soft attributes (general), 20
Soft attributes (robot-specific), 21
Soft decision trees, 12, 32
S-RL toolbox, 75
Standards, 8
Stanford, 1
StarCraft, 37, 79
State transformation
  abstract MDP, 19
  clustering states for abstraction, 24, 27, 39–40, 80
  dimension reduction or abstract MDP, 24, 27, 39–40, 52, 53, 61, 70, 80, 86
  planning over abstract graph, 24, 27, 40–44, 51
Structured causal model (SCM), 37–38, 58, 79
Subjective attributes, 16
Summary, 44–46, 71, 82
Summary, policy, 12
Superhuman AI, 75
Supervised learning, 7, 33
Swarm, 36
Symbolic methods, 11, 19
Symbolic model or policy, 24, 27, 39, 41, 52–53, 65–66, 70, 75, 84, 87
Symbolic policy by distillation, 24, 27, 64–65, 78
Symbolic rewards by evolution, 24, 27, 65, 87

T

Teams, human-robot, 8
Theory of mind, 13, 73, 76
Time required to process, 15
Trajectories, 31, 45, 81
Transferable augmented instruction graphs (TAIG), 52, 80
Transfer learning, 3, 52, 62, 70, 76, 84
Transparency, 8, 15
Trust, 3, 8
T-SNE, 40

U

Uncertain model reconciliation, 25, 28, 57, 85
Unmanned aerial vehicles (UAV), 33, 34, 38, 48, 69
Unsupervised learning, 7, 43
Use custom domain specific language as action components, 24, 49
User study, 33, 45, 46, 69–72, 78, 86
  discussion of, 12
  explanations, 8
  human-robot teams, 8
  trust, 8
Using RL to learn DT by additive process, 24, 27, 30–32
UTree, 30

V

Validation, 3, 83, 86
Value alignment, 74
Variational autoencoder (VAE), 39, 43–44, 54, 72, 81
Verification, 8, 12
Visual entity graphs, 42, 80–81
Visualizations, 11, 12, 31, 33, 34, 38, 61, 86
VRKitchen, 75

W

Watch and help environment, 76
When-produced, 4, 14, 18–19, 23, 85
  after, 17, 24, 26, 30, 32, 33, 35–37, 39, 47, 48, 55–58, 60, 64
  before, 17, 24, 26, 49, 62
  during-byproduct, 17, 24, 26, 34, 36, 37, 40, 45, 46, 50, 58–60, 65
  during-intrinsic, 17, 24, 26, 30, 39, 40, 52, 54, 58, 65, 66

SPRINGER NATURE

GPSR Compliance

The European Union's (EU) General Product Safety Regulation (GPSR) is a set of rules that requires consumer products to be safe and our obligations to ensure this.

If you have any concerns about our products, you can contact us on ProductSafety@springernature.com

In case Publisher is established outside the EU, the EU authorized representative is:

Springer Nature Customer Service Center GmbH
Europaplatz 3
69115 Heidelberg, Germany

The manufacturer's authorised representative in the EU is Springer Nature Customer Service Centre GmbH, Europaplatz 3, 69115 Heidelberg, Germany. If you have any concerns regarding our products, please contact ProductSafety@springernature.com

Printed and bound by CPI Group (UK) Ltd, Croydon, CR0 4YY

25/03/2026

02078171-0017